The Scrolls That Burned

The Scrolls That Burned

The Rise and Fall
of the Library of Alexandria

Charles Kensington

Hidden Histories
Media

Staunton, VA 24401

Printed in the United States of America

First Printing, 2025

Published by Hidden Histories Media

Paperback ISBN: 978-1-971137-18-6

Contents

Introduction

A narrow beam of light falls across a stone floor as a door opens into a room that has been closed for years. Dust stirs in the air, rising slowly as someone steps forward and pauses just inside the threshold. The smell of dry paper and age hangs thick, untouched by movement for a long time. Shelves line the walls, their shapes barely visible beneath layers of shadow. On a table near the center of the room, a bundle of old pages rests exactly where it was left, tied with a thin strip of cloth that has grown brittle with time.

A hand reaches out and lifts the bundle with care. The cloth gives way with a soft sound, and the top page is eased open. The writing is faint, its lines broken in places where the surface has worn thin. Yet the words remain. They have waited here, silent and still, through years when no one came to read them. The room holds its breath, as if the act of opening the page has stirred something that has long been at rest.

Moments like this happen in quiet places across the world. A forgotten shelf in a monastery, a sealed chest in a library, a stack of papers tucked away in a corner room. The discovery is rarely dramatic. There is no sudden noise, no grand announcement. Only the steady, careful movement of someone who understands that what they hold has passed through many hands before reaching this moment. The past returns in silence, carried in fragile ink and fading fibers.

Yet each time a long-hidden text is uncovered, a question rises with it. If these pages were kept so carefully, why were they set aside in the first place? Why were they moved from open shelves into locked rooms, from busy halls into quiet storage? And what made people choose to protect them, even when they were no longer used?

Across centuries and cultures, the same pattern appears. Knowledge is gathered, studied, and shared. It becomes part of daily life, shaping how people think and speak. Then, slowly, something changes. Access narrows. A text is handled with greater care. A door is closed, not forever, but long enough for memory to fade. The words are not destroyed. They are preserved. They are contained.

This book begins with a place that became the most powerful symbol of that pattern. In one city, long ago, people brought together writings from many lands. They filled rooms with scrolls and pages, building a center where knowledge could be gathered and studied. The work was careful and steady, shaped by the belief that what was written mattered and should be protected. For a time, those rooms stood open, alive with voices and movement.

But the story did not end there. The rooms did not remain full forever. Fires came. Changes in power came. Decisions were made that shifted how the writings were handled. Some were lost. Others were carried away. Many were copied and passed on. The place that once held them became a memory, spoken of in fragments and remembered in pieces.

And yet, even after the walls fell silent, the pattern continued. In other cities and other times, people faced the same choices. They gathered texts. They studied them. Then they limited who could read them. They moved them into storage. They sealed them behind doors. Not always out of fear. Sometimes out of caution. Sometimes out of care.

The question at the heart of this book grows from those moments. What happens when knowledge becomes something to be protected rather than shared? How do people decide when to open a page and when to close it? And what remains when the work of reading and teaching slows, leaving the words to rest in silence?

There is something deeply human in the way these choices unfold. A teacher copies a fading line so it will not vanish. A scholar wraps a fragile page and sets it on a shelf. A caretaker records where a text has been placed and who last held it. Each act is small. Each decision seems quiet at the time. Yet together, they shape the path knowledge takes as it moves across years and across borders.

The story of lost books is not only about fire and destruction. It is also about the slow changes that happen when people choose to step back. A writing might remain safe in a chest for decades. It might be opened again by someone who does not know the full story of how it came to rest there. It might be copied once more and then set aside again. The cycle repeats, carried forward by steady hands and careful choices.

In many places, the pages that survived were not the ones read most often. They were the ones that were protected, moved, and stored. Their survival depended not on their use, but on the decisions made around them. Someone chose to keep them. Someone chose not to throw them away. Someone chose to let them wait.

That waiting is part of their story. A text might rest in silence for years, then return to light in a moment of discovery. When it does, it brings with it the marks of all the hands that cared for it. Faded ink. Worn edges. Small notes written in the margins. Each detail shows that it was held, handled, and protected across time.

The journey this book takes begins in one city and moves outward. It follows the rise of a place that gathered knowledge from many lands and the moments that led to its change. It traces how writings were copied, carried, and preserved after the halls that once held them were gone. It looks at the choices made by those who guarded them, sometimes openly, sometimes in silence.

The path leads through rooms where teachers worked in small groups, through monasteries where texts were stored in chests, through libraries where certain shelves were set apart. It moves across centuries, following the same pattern as it appears again and again. In each place, the details are different. The names and settings change. But the quiet work of preserving, restricting, and protecting remains the same.

At the center of it all are the people who stood between the page and the world. The scribes who copied fading lines. The teachers who spoke words from memory. The caretakers who closed doors and kept lists. Their actions shaped what survived and what slipped away. They rarely left grand records of their choices. Often, the only trace of their work is a name written in the margin or a note that a text was moved from one room to another.

The emotional pull of these stories lies in what they suggest as much as in what they show. A room filled with shelves that once held many scrolls. A chest that remains unopened for years. A page that carries the marks of careful repair. These moments speak of effort, caution, and the steady passage of time. They hint at the tension between the desire to share knowledge and the need to protect it.

There is also a sense of wonder in the survival itself. Against all odds, some texts endure. They travel through fire, change, and silence. They are copied by hands that will never be known. They rest in places where no one looks for years. And then, one day, they are opened again.

The story that follows is not a search for hidden secrets. It is a journey through what can be traced and what has been preserved. It follows the documented paths of texts that were gathered, lost, copied, and protected. It listens for the echoes of voices that once filled rooms with learning. It watches how institutions responded when knowledge became something to be handled with care.

Each chapter steps into a different moment along that path. A room where pages were first written. A place where they were copied again. A setting where they were set aside. The focus stays on the people who stood close to the work and the decisions they made. Their choices, quiet as they were, shaped how knowledge moved from one generation to the next.

The question that opened in that dusty room at the beginning remains. Why do some texts wait in silence while others are read and shared? What makes people protect knowledge rather than spread it? And what happens when the doors close, not in anger or fear, but in careful restraint?

The answers do not come all at once. They appear in fragments, in records, in the physical traces left behind. They appear in the spaces between what is known and what can only be suggested by the survival of a page. The journey ahead follows those traces, step by step, through time and across places where knowledge was gathered and then held in quiet care.

The door to the room is still open. The page lies in the light. The words that waited so long to be seen are there, ready to be read again. And beyond that moment, the story continues, leading into the first chapter, where the rise of a great center of learning begins to take shape.

Chapter 1
The Harbor That Gathered Scrolls — The Founding of the Royal Library

Before dawn along the edge of the Mediterranean, the harbor of a young and rising city begins to stir. Lantern light moves across wet wooden decks as sailors tie ropes and lift cargo from ships that have crossed long distances. Crates of grain and jars of oil are passed down the lines, but officials are watching for something else. Scribes stand ready with wax tablets and ink, noting the presence of scrolls brought from distant ports. Each bundle is carried away before the sun fully rises, handled with care and quiet urgency.

Beyond the harbor, the city stretches inland in pale stone and wide streets that are still empty in the early light. At the center stands a growing complex of halls and courtyards where attendants unroll new texts across long tables. Fragile sheets of papyrus are smoothed flat as hands begin the steady work of copying. The air smells of salt, oil, and fresh ink. Knowledge is arriving piece by piece, carried across water and placed into waiting hands.

A City Built for Power and Learning

In the years after the city was founded, its rulers worked to shape it into more than a place of trade. Wide streets, strong walls, and grand public buildings rose near the harbor, planned with care and purpose. Ptolemy I, who ruled after the death of Alexander, ordered the creation of a royal quarter where learning could stand beside government and ceremony. Within this protected district, halls were built where scholars could live, study, and work under royal support.

His son, Ptolemy II, expanded these efforts, bringing teachers, writers, and thinkers into the city from distant regions. They were given rooms, food, and a steady place to work, all paid for by the crown. Within the Mouseion, long corridors and shaded courtyards filled with quiet voices as scholars gathered to read and record. The growing library became part of this royal project, shaped by those who believed that gathering written knowledge would strengthen the city's standing in the world.

The Making of the First Collected Scrolls

In rooms set apart from the noise of the harbor, workers began preparing the materials needed to hold written knowledge. Papyrus sheets arrived in bundles from the Nile marshes, pressed flat and cut into long strips. Scribes joined the strips together with paste to form smooth surfaces that could carry many lines of writing. Ink was mixed in small bowls, and reed pens were trimmed to fine points. Tables stood in rows beneath narrow windows, where light fell across the work as hands moved steadily from morning into night.

Some texts were brought in from distant cities, already written and worn from use. Others were copied from older versions that scholars carried with them when they arrived in Alexandria. Teams of scribes read aloud as they worked, checking each line before writing it down again. Once finished, the scrolls were rolled, tied with cord, and marked with small tags so they could be placed on shelves in careful order.

Word of the Growing Collection Spreads

Merchants, sailors, and visiting scholars began to notice the steady removal of scrolls from ships entering the harbor. Some spoke in low voices as officials searched cargo for written works, recording each item before it was carried away. Travelers arriving from other cities watched scribes handle their texts with careful hands, making copies before returning them. The process drew attention, and word spread that the rulers of Alexandria were gathering written knowledge from across the sea.

Within the city, teachers and students spoke of the growing rooms filled with scrolls, each shelf holding more than the last. Some felt pride in seeing such effort placed into learning, while others wondered how such a large collection could be guarded. A few visitors hesitated before handing over their writings, unsure if they would be returned in the same condition. Even among those who worked inside the halls, there were quiet moments of concern about the strain placed on so many fragile texts.

Questions Raised in Quiet Corridors

As the collection grew, the work drew closer attention from those in positions of authority. Officials who managed the harbor kept careful watch over the steady flow of scrolls being taken for copying. Some wondered how long such a process could continue without causing tension among travelers and traders who depended on the safe return of their goods. Records had to be checked, stored, and guarded, and the growing stacks of texts required more space and careful handling with each passing season.

Among teachers and temple leaders, quiet conversations began about the influence of so many writings gathered in one place. Some welcomed the effort, seeing it as a way to preserve knowledge that might otherwise be lost. Others spoke more cautiously, unsure how such a large store of ideas from many regions would shape the thinking of those who studied them. In shaded corners and narrow hallways, these concerns were shared in low voices, without clear answers.

Rules Formed Around the Written Word

As the number of scrolls increased, scholars and officials began setting careful routines for how each text would be handled. Scribes worked in pairs, one reading aloud while the other copied each line, checking for errors before moving on. Finished scrolls were marked and stored in rooms set aside for specific subjects. Attendants were told to limit who could handle certain works, especially older texts that showed signs of wear. Records from the period describe the steady effort to keep order through written lists and careful tracking.

Later writers claimed that some texts were held back or copied without being returned, though these reports were written years after the events and cannot always be confirmed. What can be shown is that the rulers placed strong value on keeping copies within the city. Agents were sent to gather writings, and scribes were kept busy copying them as quickly as possible. Control came through routine and oversight rather than open restriction.

The Weight Carried by Hands and Minds

For the scribes who spent their days copying line after line, the work demanded steady focus and long hours. Fingers stained with ink moved across papyrus from early morning until lamplight filled the rooms. A single mistake could mean starting again, and the pressure to keep pace with new arrivals rarely eased.

Some spoke of sore hands and tired eyes, yet the work continued without pause. Those trusted with rare texts felt the strain of knowing that any damage could bring blame or the loss of position.

Scholars who had come from distant lands found their lives tied to the growing halls of learning. Many lived close to the Mouseion, separated from their families and former homes. Some gained respect for their skill and knowledge, while others struggled to find their place among strong voices and firm opinions. Quiet rivalries formed as names became known, and the effort to protect one's work and standing weighed heavily on those who stayed.

Orders Carried Through the Halls

As the collection expanded, the royal court put firm routines in place to manage the steady flow of scrolls. Officials at the harbor were given written instructions to record each text brought into the city and send it for copying before it was returned. Inside the library complex, attendants followed strict patterns for storing and tracking the growing archive. Lists were kept on tablets and sheets, noting where each scroll belonged and who was allowed to handle it.

Rooms were assigned for different kinds of writings, and some areas were set aside for older or fragile works that required careful attention. Access was shaped by role and duty, with certain texts kept under closer watch. Over time, more workers were added to keep records, carry bundles, and maintain order within the halls. The system moved steadily, guided by written instructions and quiet supervision rather than public display.

A Reputation That Traveled Across the Sea

In the years that followed, travelers began speaking of Alexandria as a place where written knowledge was gathered in great numbers. Scholars arrived from distant cities, drawn by the promise of study and access to texts they could not find elsewhere. Ships carried word of the growing collection from one port to another, and the name of the city began to stand for learning as much as trade. With each passing decade, the effort to gather and protect scrolls shaped how other centers viewed the city and its rulers.

Yet quiet concerns lingered beneath the growing pride. The work of copying and storing so many fragile texts demanded constant care, and some wondered what would happen if support ever faded or conflict reached the harbor. A few older scribes spoke of how easily papyrus could be damaged by time, damp air, or careless handling. Even as the halls filled, there was a sense that the safety of so much knowledge depended on steady attention that might not always hold.

Fragments That Still Speak

Centuries later, only scattered traces remain to show how the early collection took shape. Bits of papyrus found in dry ground, worn tags once tied to scrolls, and references written by later scholars offer small glimpses of the work that filled the halls.

Archaeologists have uncovered pieces of buildings in the royal quarter, along with tools used for writing and storage. These finds, though incomplete, point to the scale of effort required to gather and copy so many texts in one place.

In archives and museums far from the coast, careful hands now preserve the surviving fragments. Scholars handle them under soft light, turning each piece slowly to read faint lines left by long-gone scribes. Some writings may trace back to copies first made in those early years, carried across regions as time passed. The harbor where the work began still faces the sea, its waters moving as they did when the first scrolls were brought ashore.

Chapter 2
The Shelves That Multiplied — Growth of a Library Without Limits

Morning light filters through high windows and falls across rows of newly built shelves that stretch deeper into the halls than before. The sound of pens moving over papyrus fills the air as scribes sit shoulder to shoulder at long tables. Lamps still burn from the night's work, their smoke drifting toward the ceiling as attendants carry in fresh bundles of scrolls wrapped in cloth. Each arrival is placed carefully on waiting surfaces, unrolled just enough to read the first lines, then set aside for copying.

In the corridors, footsteps echo as workers move from room to room, stacking finished scrolls in tall cases that rise above eye level. Outside, the harbor remains crowded with ships, their cargo passing quickly into the city. Inside, the pace never slows. Shelves are added where space allows, and new rooms open to hold what keeps arriving. The growing collection fills the building with a steady sense of motion, as if the walls themselves are expanding to keep up.

A Royal Project Expands Its Reach

As the city continued to grow under the rule of Ptolemy II, the royal quarter became busier with each passing year. Courtyards that once held small groups of scholars now filled with workers carrying bundles of papyrus and tablets from one hall to another. New rooms were added beside older ones, built to hold the rising number of scrolls gathered from distant ports. Officials walked the corridors with tablets in hand, keeping careful records of where each new space would serve the growing work.

Within the Mouseion, daily life followed a steady rhythm shaped by royal support. Scholars were given rooms to live and study, and scribes were assigned places where copying could continue without pause. Supplies arrived regularly from the Nile and nearby workshops, ensuring that papyrus, ink, and tools were always ready. The collection had moved beyond its earliest stage, becoming part of the city's routine as workers and scholars passed through the halls from morning until night.

Hands at Work on Endless Copies

Inside the long workrooms, fresh sheets of papyrus were laid out in neat rows, their pale surfaces waiting for lines of ink. Workers brought in bundles cut from reeds along the Nile, pressed and dried until they formed smooth strips ready to be joined. Scribes trimmed their pens and mixed ink in shallow bowls, testing the flow on scraps before beginning the careful work of copying. Some texts arrived worn and faded from years of use, and each line had to be read slowly to prevent errors.

Teams worked side by side, one voice reading aloud while others wrote in steady rhythm. When a copy was finished, it was left to dry before being rolled and tied with thin cords. Small tags were attached to mark the author or subject so attendants could carry the scroll to its proper place. The rooms filled with the soft scratch of pens and the smell of papyrus dust as new texts were created from older ones.

Rumors Carried Through the Port

Sailors and merchants began to notice that the work inside the halls never seemed to slow. As they moved through the harbor, they watched attendants carry new bundles of scrolls past the gates and into the growing complex. Some spoke of rooms filled wall to wall with writing, a sight few had ever seen. Visitors who passed through the city told others that texts from many lands were being copied and stored in numbers that were hard to imagine.

Among travelers, there was both interest and quiet caution. A few hesitated when officials asked to see the writings they carried, unsure how long the copying might take or how carefully their scrolls would be handled. Teachers who arrived hoping to study the collection spoke in low voices about how quickly it was expanding. Some wondered how the workers could keep track of so many texts without losing order as the shelves continued to fill.

Unease Behind Closed Doors

As the halls filled with more scrolls than ever before, officials began to speak quietly about the strain on space and order. Storage rooms that once held a few hundred texts were now crowded with tightly packed bundles. Workers were asked to move materials from one room to another to make space for new arrivals. The growing collection demanded more hands, more supplies, and closer watching, and some administrators worried about how long the system could keep up with the pace.

Outside the library complex, temple leaders and teachers spoke in low tones about the rising influence of so many writings gathered in one place. Texts from distant regions carried unfamiliar ideas and beliefs, and not everyone welcomed their presence. Some questioned whether every scroll should be copied and kept, while others urged patience and care. In meeting rooms and shaded courtyards, these concerns lingered, shared in cautious voices as the shelves continued to fill.

Quiet Rules for Order and Access

As the number of scrolls increased, scholars began setting careful routines for how each text would be handled and placed. Lists were written to track titles and authors, and attendants were instructed to carry finished copies to assigned rooms. Scribes were told to work from clear versions of each text to prevent errors from spreading. Records from the period describe the steady effort to keep the growing archive in order through written notes, marked tags, and careful storage.

Later accounts claimed that some writings were kept under closer watch or copied more than once to protect their contents, though these reports were written years after the early growth of the collection and cannot always be confirmed. What can be shown is that certain rooms were set aside for older or fragile works, and access was shaped by the duties of those who worked inside. The routines formed slowly, guided by the need to protect what had been gathered.

The Toll of Endless Work

For the scribes, the steady rise in copying brought longer hours and growing strain. They bent over tables from early morning until late into the night, their hands darkened by ink and their backs stiff from leaning forward. A single mistake could mean losing days of work, and the pressure to keep pace with new arrivals never seemed to ease. Some began to avoid crowded rooms to protect their focus, choosing quieter corners where they could work without interruption.

Scholars who lived within the complex felt a different kind of weight. Their names were tied to the accuracy of the texts they handled, and quiet rivalries formed as the collection grew. A missed line or damaged scroll could harm a reputation built over years. Many had left their homes to live close to the halls, separated from families and familiar places. In narrow rooms lit by oil lamps, the work often felt endless, and the silence carried its own kind of strain.

New Routines Set Into Place

As the collection continued to grow, officials began adding more structure to how the work was carried out. Written orders were sent to the harbor, reminding inspectors to record every text brought into the city and send it forward for copying. Inside the complex, attendants followed fixed paths, carrying bundles from receiving rooms to copying tables and then to storage areas. Shelves were rearranged to make room for new scrolls, and some rooms were set aside for texts that needed special care.

Over time, additional workers were assigned to keep records and track movement within the halls. Scrolls were marked with small tags so they could be returned to their proper place after use. Some writings were moved between rooms as space grew tight, and access was shaped by the duties of those who worked inside. The process moved quietly, guided by written instructions and steady supervision.

A Collection That Outgrew Its Walls

In the decades that followed, the steady work of copying and storing changed the shape of the complex itself. New rooms were added where courtyards once stood, and older spaces were filled from floor to ceiling with scrolls tied in bundles. The city began to be known for the size of its collection, and travelers spoke of the halls where knowledge seemed to multiply with each passing year. Scholars arrived hoping to find texts they had only heard about, and movement through the corridors rarely slowed.

Yet the growing scale brought quiet uncertainty. Some wondered how the workers would continue to track so many writings without losing their place. Others spoke in low voices about the strain on shelves, rooms, and hands already burdened by constant use. The more the collection expanded, the more it depended on careful attention and steady support, and no one could be sure how long that balance would hold.

Marks Left by the Growing Halls

Today, small pieces of evidence offer a sense of how large the collection once became. Fragments of papyrus, recovered from dry ground and burial sites, still show the careful lines written by trained hands. Some carry notes added by scribes who copied them, marking titles or authors in the margins. Archaeologists have found parts of storage rooms and writing tools that hint at the scale of the work, even though the full complex no longer stands.

In climate-controlled archives, these fragile pieces are handled with great care. Curators turn each fragment slowly, reading faint letters that have survived through centuries of movement and change. A few texts may trace back to copies made during those years of rapid growth, passed from one owner to another as time moved on. The marks of ink and the smooth touch of papyrus still carry the quiet record of hands that worked without rest.

Chapter 3
The Voices Within the Walls — Scholars, Translators, and the Work of Preservation

Late in the afternoon, soft light falls across a shaded courtyard where voices rise and fade in steady rhythm. Small groups sit beneath stone columns, unrolling scrolls across worn tables. One man reads aloud in Greek, his voice low and measured, while another listens and follows the lines with careful attention. Nearby, a second voice speaks in a different tongue, repeating the same passage slowly, weighing each word before it is written again. The air is warm and still, carrying the faint scent of ink and papyrus.

Inside the nearby rooms, translators bend over long tables, comparing lines from one language to another. Sheets are spread wide, marked with small notes in the margins as hands move back and forth between texts. Lamps are lit one by one as the sun lowers, their glow reaching across the pages. The sound of careful speech and steady writing fills the halls, echoing through corridors where knowledge passes from voice to voice.

A Meeting Place for Many Tongues

Within the royal quarter, the Mouseion had become a place where scholars from distant regions lived and worked side by side. Rooms opened into courtyards where voices carried in many languages, and the daily rhythm of study filled the air from morning until night. Under the support of the Ptolemaic rulers, teachers, readers, and writers were given food, lodging, and space to continue their work. The growing library stood close by, offering access to texts that few had ever seen gathered in one place.

Visitors arrived from across the eastern Mediterranean and beyond, bringing with them their learning and their speech. Some came from cities known for philosophy, others from places where medicine, mathematics, or history had long been studied. In shared rooms and shaded halls, they read from scrolls brought into the collection, working together to understand writings that came from many lands and traditions.

Lines Written Across Languages

In the workrooms near the courtyard, long tables were covered with scrolls opened side by side. Some were older texts brought from distant cities, their edges worn and darkened from years of use. Others were fresh copies made on newly pressed papyrus.

Translators sat in pairs, one reading aloud from the original while another wrote each line again in Greek. Pens moved slowly, pausing often as the reader repeated a word to ensure the meaning was clear.

Sheets of papyrus were prepared in stacks, trimmed and joined to create long surfaces for writing. Ink was mixed again and again as the work stretched across the day. When a section was complete, the new text was read aloud to check for mistakes before being rolled and tied. Some translations required several versions before the final copy was placed among the growing collection, marked so it could be found and used by others.

Voices Noticed in the Corridors

Students and visitors who passed through the halls began to notice the steady exchange of languages carried from one room to another. They paused near doorways, listening as the same passage was read aloud more than once, first in one tongue and then in another. Some leaned closer to the tables where translators worked, watching as new words were written beside older lines. The sound of repeated reading filled the corridors, and the work drew quiet attention from those who had come only to study.

Among the scholars themselves, reactions formed slowly. A few spoke with excitement about the chance to bring distant writings into a language more could understand. Others showed hesitation, aware that a single word could shift meaning if not chosen with care. Some paused over certain passages, uncertain how to carry their sense across without losing what the original writer meant. In those moments, the rooms grew still as voices lowered and hands hovered above the page.

Questions Shared in Quiet Circles

As more translations appeared, some teachers and elders began to watch the work more closely. Writings from distant lands carried ideas that were not always familiar, and there were moments when a passage caused low voices to rise in careful discussion. A few scholars hesitated before reading certain lines aloud, aware that different beliefs and customs were now being spoken within the same walls. The work continued, but the air sometimes held a sense of caution.

Temple leaders and visiting officials also took notice of how widely these texts were being shared. In shaded corners and along the walkways, small groups gathered to speak in private. Some wondered how far these translated writings would spread and how they might shape the thinking of those who studied them. The effort to bring many voices together in one place began to carry a quiet pressure that was felt in the pauses between words.

Careful Hands and Measured Words

As the number of translated texts grew, scholars began setting clear routines for how the work should be checked and reviewed. Pairs of readers compared new copies to the original scrolls, reading each line aloud before it was approved for storage. Some passages were written again more than once to settle on the most accurate wording. Records from the period show that lists were kept to track which texts had been translated and who had worked on them.

Later writers claimed that certain works were handled with extra caution or placed in rooms where fewer people could reach them, though these reports come from sources written long after the early translation work began. What can be shown is that some scrolls were marked for careful study, and access was shaped by the duties of the scholars assigned to them. The effort to keep order moved quietly through repeated checking, steady reading, and patient copying.

The Weight of Every Word

For the translators, the work carried a steady strain that few outside the rooms could see. A single word could change the sense of a passage, and long hours were spent reading and rereading lines to avoid mistakes. Some grew tired from the effort of holding two languages in mind at once, their eyes sore from staring at inked letters by lamplight. When disagreements arose over how to carry meaning from one tongue to another, the tension lingered in the silence that followed.

Reputation rested on careful work, and a poorly chosen phrase could draw quiet criticism from others. Some scholars withdrew to smaller rooms to focus, choosing isolation over the risk of error in crowded spaces. Others found themselves caught between strong voices with different views, unsure which path to follow. The pressure stayed close, carried in the careful movement of each pen across the page.

Orders That Guided the Work

As translation efforts grew, the officials overseeing the Mouseion began setting clearer routines to manage the flow of texts between rooms. Lists were updated to note which scrolls had been translated and which still required attention. Attendants were directed to carry originals and copies along set paths, moving them from study tables to storage spaces once the work was checked. Some texts were placed in areas reserved for active use, while others were kept in quieter rooms where they would be handled less often.

Over time, certain duties were assigned to specific scholars, shaping who worked on particular writings. Requests to view or copy texts were passed through the hands of those responsible for keeping records. Scrolls were sometimes moved from one room to another as space was rearranged to hold new work. The process unfolded through written instructions, steady supervision, and careful tracking rather than public notice.

Words That Traveled Beyond the Walls

In the years that followed, copies made in those rooms began to appear in other cities, carried by scholars who left Alexandria to teach or study elsewhere. Writings once spoken in a single language could now be read in another, passed from hand to hand in places far from the harbor. Students who had trained within the Mouseion carried their learning with them, repeating lessons and sharing texts that had been shaped through careful work inside the library's halls.

Yet quiet questions remained about how much of the original meaning had been carried forward. Some later readers noticed differences between older texts and the copies that reached them years later. A few spoke of words that felt changed or uncertain, though no clear record could settle every difference. As time passed, the work of translation continued to spread, and the path each text had taken grew harder to trace.

Fragments That Still Carry Many Voices

Today, some of the earliest translated texts survive only in worn fragments kept in quiet archives. Pieces of papyrus show lines written in more than one language, the careful work of scribes still visible in the shape of each letter. Scholars handle these remains under soft light, studying the marks left by hands that worked long ago. In a few cases, later copies preserved in distant regions offer clues to the earlier versions once held in Alexandria.

Museums and libraries now keep these pieces in controlled rooms where heat and light are limited. Each fragment is placed between protective layers, and every movement is done with steady care. The faint lines of ink remain, showing the effort to carry meaning across languages and generations. In these silent rooms, the sound of turning pages has replaced the voices that once filled the halls.

Chapter 4
The Catalog That Mapped Knowledge — Order Inside the Growing Collection

In a quiet room set apart from the busiest halls, long wooden tables are covered with scrolls waiting to be recorded. A single lamp burns low, casting warm light across the smooth surface of papyrus as a scholar unrolls each text just far enough to read its opening lines. He pauses, listening to the faint sounds of movement in the corridor, then dips his pen and begins to write careful notes on a fresh sheet laid beside him. Assistants step in and out, placing new bundles within reach before slipping away again.

Outside the room, voices echo and footsteps pass in steady rhythm, but inside there is only the soft scratch of ink against papyrus. Scroll after scroll is opened, checked, and marked with small tags tied to the cord at one end. Each entry brings a growing sense of order to the swelling collection. The work moves slowly, guided by patience and steady hands, as knowledge is given a place where it can be found again.

Rooms Set Aside for Careful Record

Within the royal quarter, certain rooms were kept quieter than the rest, used for the steady task of recording what the library held. Shelves stood packed with scrolls tied in bundles, and tables were cleared to make space for those brought in to be listed and placed. Scholars chosen for their careful hands worked in these spaces, reading the first lines of each text and writing short notes that would guide others later. The work moved at a measured pace, shaped by routine and attention.

The Mouseion provided the setting for this effort, with scholars living and working close to the growing collection. Supplies were carried in each morning, and attendants moved through the corridors with armfuls of scrolls waiting to be sorted. The task of keeping track of the collection became part of the daily rhythm, carried out in quiet rooms while voices and footsteps continued beyond the doors.

Lists Formed from Ink and Patience

Sheets of fresh papyrus were prepared for the work of recording titles, authors, and subjects. Scholars sat at low tables with stacks of scrolls beside them, unrolling each one just far enough to read its opening words. With careful strokes, they wrote notes onto long strips that would serve as records of what the shelves held. Ink was mixed again and again as the work stretched through the day, and small tags were tied to scrolls so they could be matched to the lists being made.

Assistants moved quietly from one table to another, carrying finished entries to be set aside for safe keeping. Some lists grew long as more texts were added, each line written in steady, even script. The process required time and patience, with hands returning to the same movements over and over. In these rooms, order began to take shape through simple tools and careful labor.

Whispered Notice of Order Taking Shape

Visitors passing through the quieter corridors began to notice the steady work taking place behind open doors. They saw scholars bending over long sheets covered with careful writing, marking titles and notes beside stacks of scrolls. Some paused to watch as tags were tied and records set aside, sensing that something new was being built from the growing collection. Students spoke of the lists in low voices, curious about how the work might help them find what they needed.

Among the scholars, reactions formed slowly as the recording began to spread. A few welcomed the order it promised, believing it would make the collection easier to use. Others hesitated, unsure how such a vast number of entries could be kept without mistake. Some wondered whether older texts might be misplaced or forgotten as new ones arrived. In the quiet spaces between tables, these concerns moved through soft conversation and careful glances.

Questions Raised by the Rising Lists

As more lists were written and stored, officials began to take a closer interest in how the work was being done. The number of entries grew each week, and the task of keeping them accurate required steady attention. Some administrators walked through the rooms, watching as scribes read and recorded titles, making sure the process followed the routines that had been set. The effort to keep order among so many scrolls brought a quiet sense of pressure to those responsible for the work.

Among scholars and teachers, there were moments of hesitation as the lists continued to grow. Some wondered how easily a mistake might spread if a title was written down the wrong way. Others spoke in low voices about the challenge of keeping track of texts from many regions and languages. In narrow corridors and shaded corners, these concerns were shared in careful tones, carried between those who worked close to the growing records.

Hands Guiding the Written Records

Scholars worked carefully to shape the growing lists into something others could follow. Titles were grouped by subject and by the names of those who had written them, and each new entry was checked against the scroll itself. Some rooms were set aside for this work so the noise of the main halls would not disturb the steady focus required. Records from the time show that librarians and assistants read the first lines of each text to confirm its place before adding it to the written lists.

Later writers claimed that certain records were kept under closer watch or limited to a small number of readers, though these reports were written long after the early cataloging began and cannot always be confirmed. What can be shown is that some lists were copied more than once to prevent loss, and that attendants followed set routines when moving scrolls to and from the shelves.

The Burden of Keeping Order

The scholars chosen to keep the records worked under steady strain. Their days were spent checking titles, reading the first lines of scrolls, and writing careful notes that others would rely on. A single mistake could cause confusion for many who came searching for a text. Some stayed late into the night, their eyes tired from reading the same lines again and again to be sure the entries were correct. The work demanded patience, and the silence of the rooms often felt heavy.

Reputation rested on steady hands and clear judgment. If a scroll was placed in the wrong spot or a title written down poorly, word could spread quickly among those who used the lists. A few scholars withdrew to quieter corners to avoid distraction, choosing to work alone where they could focus without interruption. The pressure to keep the growing records in order followed them from room to room.

Orders Written Into Daily Practice

As the catalog grew, the administrators set clear routines to guide how the work would continue. Written instructions were passed to attendants, directing them to bring scrolls to the recording rooms in steady order rather than all at once. Lists were kept near the entrances of storage areas so workers could check each title before placing a text on the shelves. Scribes followed the same steps each day, reading, noting, and returning scrolls to their assigned places.

Over time, certain records were moved to safer rooms where fewer hands would reach them. Copies were made of the longer lists so that if one was damaged, another would remain. Access to some areas was shaped by the duties of those assigned to the work. The movement of texts, tags, and written entries continued under quiet supervision, guided by routine rather than open display.

Paths Made Through Written Order

Years later, the lists created in those quiet rooms shaped how scholars moved through the growing collection. New arrivals learned to search by the names and subjects recorded long before their time. The written entries guided attendants as they carried scrolls from one shelf to another, helping them return each text to its place. Over decades, the habit of recording and sorting became part of the daily life of the halls, passed from one group of workers to the next.

Yet there were moments of uncertainty that stayed with those who relied on the records. Some older entries faded or were damaged as years passed, and a few titles became harder to trace when scrolls were moved or replaced. Scholars spoke in low voices about the risk of losing track if the lists were not carefully kept. Even with careful work, no one could be sure every record would remain clear as time moved on.

Lines That Outlasted the Shelves

Today, only scattered signs remain of the early lists that once guided scholars through the halls. Small fragments of papyrus, found in dry ground and hidden places, still show rows of careful writing that may have been part of catalog records. Some carry faint marks where titles were noted, the ink worn but still visible under gentle light. These pieces offer a quiet glimpse into the work of those who tried to bring order to a growing sea of texts.

In archives far from the ancient city, such fragments are kept between protective layers, handled with slow and steady hands. Researchers study the spacing of the lines and the style of the script, tracing patterns that hint at how the records were made and used. The lists themselves are gone, yet the careful writing that once guided others can still be seen in these surviving traces.

Chapter 5
The King's Command – Gathering Scrolls From Every Shore

At the harbor, a ship from a distant port rocks gently against the stone dock, its sails still damp from the long journey. Dockworkers pull thick ropes tight and lower gangplanks as officials step aboard carrying tablets and writing tools. Crates are opened one by one, lids lifted to reveal trade goods wrapped in cloth and sealed jars stacked close together. Among them, bundles of scrolls lie tucked between cargo, some worn from travel, others bound tightly with cord.

Attendants waiting on the dock take each scroll with steady hands, noting its arrival before carrying it away through the gates. Sailors watch as the texts are removed, their eyes following the movement toward the streets that lead inland. Beyond the noise of the port, runners hurry toward the royal quarter where rooms stand ready to receive new writings. The work moves quickly, guided by routine and quiet authority, as the flow of ships brings more words from distant shores.

Orders Sent Across the Harbor

Under the rule of the early Ptolemies, the harbor became more than a place of trade. Officials were placed at the docks to watch the steady flow of ships entering the city. They carried writing tablets and kept careful records of goods and travelers arriving from distant ports. Among the crates and bundles, they looked closely for written works that had come across the sea. The gathering of texts was treated as part of the city's daily business.

Within the royal quarter, instructions moved quietly from one office to another. Attendants were told to bring newly found scrolls into the library complex, where scribes waited to copy them. Workers carried bundles through narrow streets, moving quickly but with care. The effort to gather writing from many regions became part of the routine of the harbor, shaped by written orders and steady supervision from those close to the throne.

Scrolls Carried from Distant Shores

Bundles of papyrus arrived wrapped in cloth or packed inside wooden chests, protected from damp air and rough travel. Some scrolls were written on fresh sheets with dark, steady lines, while others showed signs of age, their edges worn and surfaces marked by many hands. Sailors carried them down from the ships along with trade goods, placing them on the dock where officials checked their contents. Each text was lifted carefully, its cord untied just enough to see the first lines and confirm what it held.

Inside the workrooms, scribes prepared new sheets of papyrus to copy what had arrived. Ink was mixed, and pens were trimmed before the work began. Some texts had been written in distant cities, shaped by local customs and materials. Now they were laid open on tables where teams of workers read and wrote in steady rhythm, turning older words into new copies that could be placed among the growing collection.

Eyes Watching at the Water's Edge

Sailors and merchants began to notice how often their ships were searched when they entered the harbor. Officials moved with quiet purpose, lifting bundles and untying cords to check for scrolls among the cargo. Some travelers stood close by, watching as their writings were carried away by attendants. Word spread along the docks that the rulers of the city were gathering texts from every shore. The sight of scribes waiting near the gates became a familiar part of arrival.

Among the crews, reactions formed in low conversation. A few showed pride in knowing their writings would be copied and kept in a place known for learning. Others hesitated as they handed over their scrolls, unsure how long the process would take or how carefully their work would be treated. In the crowded noise of the harbor, these concerns lingered in quiet whispers and careful glances.

Uneasy Talk Along the Quays

As the practice of collecting scrolls became more common, officials at the harbor began speaking quietly among themselves about the strain it placed on daily work. Inspecting cargo now took longer, and ships waited at the docks while attendants searched for texts. Some administrators worried about keeping trade moving smoothly while still following the king's orders. The need to check every bundle and record each item added pressure to an already busy port.

Beyond the docks, temple leaders and teachers in the city also took notice of the growing flow of writings from distant regions. Some wondered how ideas carried from other lands might affect those who read them. In shaded walkways and narrow courtyards, these thoughts were shared in careful voices. A few questioned whether every scroll should be brought into the halls, while others urged that the gathering continue without delay.

Careful Handling Under Royal Orders

Records from the early years describe how officials followed clear steps when scrolls were brought from the ships. Each text was checked, noted, and carried into the city to be copied before it was returned or stored. Scribes worked in steady pairs, reading aloud while another wrote, checking each line as they moved forward. Attendants kept lists of what had arrived and where it was placed, following the routines set by those in charge of the collection.

Later writers claimed that some scrolls were kept in the city instead of being sent back to their owners, though these reports come from sources written long after the practice began and cannot always be confirmed. What is shown in earlier records is the steady effort to gather and copy as many writings as possible. The work was shaped by routine, careful handling, and written instructions that guided the movement of each text.

Strain Along the Dockside

For the harbor inspectors and attendants, the work brought long hours under watchful eyes. Each bundle had to be opened with care, each scroll recorded and carried away without damage. A missed item or a poorly handled text could bring sharp words from those above them. Some workers stayed at the docks from sunrise to dusk, moving between ships and storage rooms with little rest. The constant pressure to keep pace with arrivals settled into their daily routine.

Sailors and travelers also felt the weight of the practice. Some waited longer than expected while their scrolls were taken for copying, unsure when they would be returned. A few worried about the safety of their writings and spoke in low tones among themselves. In crowded taverns near the harbor, stories were shared about the inspections, and the tension followed them even after they left the city behind.

Orders That Shaped the Harbor Routine

Over time, the process at the docks settled into a steady pattern guided by written instructions. Officials were assigned fixed posts to inspect arriving ships, and attendants waited nearby to carry any scrolls that were found. Tablets were filled with notes listing what had been taken for copying, and runners moved between the harbor and the library to keep the work moving. The steps were repeated with each arrival, and the system began to feel like part of the daily rhythm of the port.

Inside the city, rooms were set aside to receive the scrolls brought from the ships. Some were placed in areas where they could be copied at once, while others waited in storage until space opened. Workers followed quiet routines, passing bundles from hand to hand along set paths. Instructions were given in writing, and the movement of texts continued without public display or open discussion.

A Practice Remembered Along the Trade Routes

In the years that followed, stories of the harbor inspections spread across the ports of the eastern Mediterranean. Sailors spoke of a city where written works were taken ashore to be copied before ships could depart. Some travelers began carrying their scrolls wrapped more carefully than before, aware that they might be handled by unfamiliar hands. The practice became part of the city's reputation, known to those who arrived long after the early years had passed.

Yet quiet uncertainty stayed with those who remembered the early days. Some wondered how many writings had been lost to time, damage, or simple misplacement during the steady flow of copying and storage. A few older workers spoke of scrolls that never seemed to return to the docks, though no clear records settled every question. The effort to gather knowledge had left a mark that stretched far beyond the harbor.

Faint Records of Ships and Scrolls

Today, only scattered traces remain to show how texts once moved from ship to shore. Fragments of papyrus found in dry burial sites still carry faint marks of titles and notes, suggesting that many copies were made from older works. A few ancient references describe harbor officials recording the arrival of goods, and these small records hint at the steady flow of writing that passed through the port. Pieces of storage jars and wooden cases have also been uncovered, offering a quiet sign of how scrolls may have been carried and protected.

In modern archives, curators handle surviving fragments with steady hands, turning each piece slowly beneath soft light. The thin layers of papyrus, marked by careful lines of ink, are placed between protective sheets and stored in controlled rooms. Some may trace their origins to copies made in those early years, carried from place to place across centuries. The memory of the harbor's work remains in these fragile traces.

Chapter 6
The First Flames at the Harbor — War Reaches the Scrolls

Smoke drifts low across the water as shouts rise along the crowded harbor. Ships packed tightly together rock against the docks while soldiers run across the planks, their armor catching flashes of firelight. Flames leap from one vessel to the next, fed by ropes, wood, and wind. The air grows thick and hot as men rush to cut lines and push boats away from the burning ships. Crates are dragged across the stone quay, and carts rattle over the ground as workers try to move what they can.

Near the edge of the waterfront, storage buildings stand close to the docks, their doors thrown open in the rush. Attendants hurry in and out, carrying armfuls of scrolls and bundles tied with cord. Ash begins to fall, drifting through the streets and settling on the roofs and courtyards beyond. The fire was meant for ships, but it spreads without pause, pushed by the wind toward the heart of the harbor.

A City Caught Between Armies

In the middle of the first century before the common era, Alexandria stood as a crowded and restless city shaped by power and rivalry. Roman forces had entered its streets during a struggle tied to the rule of Egypt, and soldiers moved through the harbor district where trade once filled the air with noise and motion. Ships from many regions remained packed along the docks, and the waterfront had become a place where tension stayed close beneath the surface.

The harbor stood near storehouses and workrooms used for trade and storage. Bundles of goods were stacked high, and records were kept of what passed through the port each day. Within walking distance, buildings connected to the collection of scrolls stood in the same busy quarter. Workers, scribes, and officials moved between these places as part of their routine, their paths crossing near the water where ships came and went.

Scrolls Stacked Near the Water

In the storage rooms close to the harbor, bundles of scrolls were stacked inside wooden chests and along narrow shelves. Many had been copied on fresh sheets of papyrus and tied with cord to keep them from unrolling. Others were older texts brought in from distant ports, their surfaces worn from years of handling. Attendants moved through these rooms each day, lifting bundles and placing them where they could be found again when needed.

Papyrus sheets came from the marshes along the Nile, pressed and dried into smooth surfaces ready for writing. Scribes worked at long tables not far from the docks, copying texts that arrived by ship. Once finished, the new scrolls were left to dry before being rolled and stored. The rooms near the water held both older writings and newer copies, placed side by side among the goods that passed through the busy port.

Fear Along the Waterfront

Workers along the harbor were among the first to notice the danger as smoke began to drift across the docks. Attendants who moved in and out of the storage rooms paused, looking toward the ships where flames had taken hold. Some rushed back inside to gather what they could carry, lifting bundles of scrolls into their arms. Others stood in the open, unsure how quickly the fire might spread or how close it would come to the buildings that held the texts.

Sailors and dockworkers called out to one another, their voices rising over the sound of burning wood and snapping ropes. A few hesitated before entering the storage rooms, uncertain if there was time to save what lay inside. Scribes who had been working nearby stepped into the streets, watching ash drift through the air. In those moments, quiet concern moved from person to person as the fire edged closer.

Orders Shouted Over the Roar

As the fire spread along the ships, officials began shouting orders across the harbor, trying to bring some control to the confusion. Guards moved people away from the docks while others were sent toward the storage buildings to see what could be carried out. The heat pushed back those who came too close, and smoke made it hard to see where the flames would reach next. Workers looked to their supervisors for direction, waiting for signs of what should be saved first.

In the streets nearby, scholars and attendants gathered in small groups, watching the rising smoke with growing worry. Some spoke of the scrolls stored near the water, unsure how many were inside or how quickly they could be moved. The noise of the harbor carried their voices as tension spread through the crowd. Pressure rose in the rush to act before the fire closed in further.

Hands Reaching Through Smoke

Workers rushed into the storage rooms closest to the water, lifting bundles of scrolls and passing them along a hurried line toward safer ground. Some tried to carry more than they could hold, their arms filled with tied rolls of papyrus as smoke thickened in the air. Guards called for the doors to be shut where the heat grew too strong, while others kept moving in and out, unsure how long the building would stand. Records from the time speak of people trying to remove texts from danger as the fire spread along the docks.

Later writers claimed that many thousands of scrolls were lost in that first blaze, though these reports were written years after the event and the exact number cannot be confirmed. What is shown in earlier accounts is the rush to protect what could be reached and the struggle to carry texts away from the smoke and heat.

Faces Marked by Smoke and Loss

In the days after the flames swept through the harbor, the workers who had rushed into the storage rooms carried the strain in their silence. Some had breathed in thick smoke while trying to lift bundles from the shelves, and their voices grew rough from the effort. Others moved slowly, their hands unsteady from the heat and fear that had followed them back into the streets. A few spoke in low tones about scrolls they had seen fall or scatter, unsure if they had done enough to save what they could reach.

Among the scribes and attendants, quiet tension settled into the halls. Those who had been near the docks felt the weight of what might have been lost under their watch. Conversations were short and careful, and some chose to work alone rather than face the questions of others. The smell of ash lingered on their clothes, and the memory of the fire stayed close as they returned to their tasks.

Records Written in the Aftermath

In the weeks that followed, officials moved through the harbor district to check the damaged buildings and take note of what remained. Workers were directed to gather any scrolls that had been carried away and place them in safer rooms farther from the water. Lists were updated as best they could be, with attendants noting which bundles had been found and where they had been moved. Some storage areas were closed off while repairs were made, and access was limited to those assigned to the work.

Over time, new routines were set to keep materials away from the most crowded parts of the docks. Scrolls were carried to rooms deeper inside the city, and guards were posted near certain entrances. The movement of texts followed quiet orders passed from one office to another. The work continued under steady watch, shaped by written records and careful handling.

Memory Carried by Those Who Remained

In the years that followed, the fire was spoken of in quiet voices by those who had seen the smoke rise over the harbor. New workers arrived who had not been there that day, but the older attendants passed along what they remembered. Some storage rooms were kept farther from the water, and a few shelves stood less crowded than before. The effort to copy and gather texts continued, though the memory of the flames stayed close to the daily work.

Even decades later, there were moments when uncertainty returned. Some wondered how many scrolls had been lost before anyone could count them. A few older records no longer matched what stood on the shelves, and no one could be sure when the change had taken place. In quiet corners of the halls, the story of the harbor fire remained part of the air, carried forward in careful conversation and uneasy silence.

Ashes and Fragments Beneath the Sand

Archaeologists working near the ancient harbor have uncovered layers of burned wood, broken pottery, and darkened soil that point to intense fire along the waterfront. In some places, pieces of papyrus have been found in brittle fragments, their edges blackened and curled from heat. These small remains are handled with care, lifted slowly from the ground and placed into protective cases. Even when the writing has faded, the shape and texture still show what once lay there.

In laboratories and archives, specialists study these fragile pieces under soft light. Each fragment is measured, cleaned, and stored where heat and air are kept steady. Some show faint lines of ink that once formed words now lost to time. The charred remains, carried through centuries, hold a quiet trace of the moment when fire moved across the harbor and reached the stored texts.

Chapter 7
The Scholar Who Guarded the Shelves — A Librarian's Final Watch

Morning light slips through the high windows and falls across rows of tightly packed scrolls. Dust drifts slowly in the air as a lone figure walks between the shelves, pausing now and then to touch a cord or straighten a bundle that has shifted out of place. His steps are quiet on the stone floor, the only sound in a room that once held many voices. In his hands, he carries a small writing tablet, marking down changes with careful strokes before moving on.

Outside, the city stirs with distant noise, but inside the halls the work continues at a slower pace. Attendants come and go, carrying new texts to be placed among the old. The man stops at the end of a row, looking over the shelves that stretch into shadow. He lingers for a moment, as if counting each bundle by memory, then turns and walks deeper into the room where the light fades.

A Quiet Steward in the Halls of Learning

In the later years of the great collection, the library remained a place where scholars lived and worked close to the shelves they cared for. Within the Mouseion, rooms were set aside for study, copying, and the keeping of records. Librarians moved through these spaces each day, checking the placement of scrolls and noting any changes. Their work was part of a steady routine that held the growing collection together.

The city around them shifted with the passing of time, but inside the halls the daily tasks continued. Attendants carried bundles from one room to another, and scribes worked at tables where light from high windows fell across fresh sheets of papyrus. Those responsible for the shelves kept watch over what had been gathered, moving through the rooms with quiet purpose as the work of preserving the collection went on.

Notes Written for the Keeper's Records

On a low table near the shelves, a librarian worked over sheets of papyrus prepared for daily records. Fresh strips were cut and smoothed, ready to hold notes about where each scroll had been placed and which had been moved for study. Ink was mixed in small bowls, and the pen was trimmed again and again to keep its tip sharp. With careful strokes, the keeper wrote short entries, marking changes that would help others find what they needed.

Bundles of older lists were kept nearby, tied with cord and stored within reach. When a new text arrived or an older one was moved, the keeper opened these records and added another line in steady script. The work required patience and steady hands, built from simple tools and repeated effort. Each entry was part of the daily task of keeping track of what the shelves held.

Quiet Notice Among the Attendants

Attendants passing through the rooms began to notice how closely the librarian watched over the shelves. They saw him pausing to check each bundle and writing careful notes about small changes. Some stopped to watch as he opened older records and added new lines in steady script. His attention to detail drew quiet respect from those who worked nearby, and they began to rely on his memory of where each scroll had been placed.

Yet there was also a sense of unease that moved through the halls. A few attendants noticed how often he returned to certain shelves, as if checking them more than once. Some wondered whether something had been misplaced or if changes had gone unnoticed. They spoke in low voices as they worked, glancing toward the rows where he stood, uncertain what had drawn his steady watch.

Questions Raised in the Inner Rooms

Senior scholars began to notice how often the keeper walked the same rows, checking cords and bundles long after others had left for the day. A few paused at the doorway to watch him work, their faces quiet as they looked over the shelves. They spoke softly among themselves, wondering if the growing collection had become harder to manage than before. The need to keep every text in its place began to feel heavier as more scrolls arrived.

Within the offices nearby, administrators asked for updates more often, requesting lists and records to be brought to them for review. Messengers carried tablets from one room to another, and the keeper was called more than once to answer careful questions. The pressure stayed close as routines were checked and rechecked, and the watch over the shelves grew more constant.

Efforts to Keep the Records in Order

The keeper spent long hours reviewing the written lists that marked where each scroll had been placed. He compared new entries with older records, tracing the movement of texts from one room to another. Attendants were asked to bring certain bundles so they could be checked against the notes on the tablets. Records from the time show that librarians relied on these lists to track the growing collection, updating them whenever changes were made.

Later writers claimed that some keepers restricted access to certain shelves to protect fragile texts, though these reports come from much later sources and cannot always be confirmed. What is shown in earlier records is the steady effort to maintain order through careful notes and repeated checks. The work continued quietly, guided by the written lists that helped shape where each scroll was kept.

A Watch That Never Truly Ended

The keeper's days grew longer as the collection expanded and the records demanded more care. He moved from shelf to shelf with little rest, checking cords, unrolling scrolls just enough to confirm their place, then writing small notes on his tablet. Attendants noticed how often he worked alone, staying behind after others had left the rooms quiet and dim. The effort to keep track of every change weighed on him, and the silence around the shelves seemed to follow him wherever he went.

Among the scholars, small signs of strain began to show. Some wondered if the work had become too much for one man to hold together. A few spoke in hushed tones about missing entries or delays in finding certain texts. The keeper heard these quiet remarks as he passed, and the pressure stayed close, carried in every careful step he took between the rows.

Written Orders for the Shelves

Over time, the offices overseeing the collection began sending clearer instructions about how the shelves were to be managed. Lists were requested more often, and the keeper was asked to submit updated records showing where each group of scrolls had been placed. Messengers carried tablets between rooms, and attendants were directed to follow set paths when moving texts from one section to another. The work began to follow written steps that guided each action.

Some areas were set aside for texts that were used less often, and certain shelves were kept closed except for those assigned to handle them. Bundles were moved deeper into the rooms where they could be watched more closely. The handling of scrolls followed quiet routines shaped by written orders, and the movement of each text was noted and passed along through the records.

Habits That Outlived the Keeper

Years after the keeper first began his careful rounds, the routines he followed remained in place. New librarians learned to walk the same paths between the shelves, checking cords and noting changes in the written records. The lists he had helped shape were still used to guide where texts were placed and how they were found. Attendants carried out the same tasks, moving quietly from one room to another as the collection continued to grow.

Yet small questions lingered in the halls. Some older entries no longer matched what stood on the shelves, and no one could be certain when the changes had begun. A few records had grown faint or damaged with time, making it harder to trace the path of certain scrolls. Those who worked among the rows felt the weight of these gaps, even as the daily routines carried on.

Marks Left in Ink and Memory

Only a few written traces remain that show how the shelves were once watched and recorded. Fragments of papyrus discovered in later centuries still carry short notes and list-like markings, lines written in a steady hand that once helped guide the movement of scrolls. Some of these pieces show repeated names and short entries, suggesting records kept close to the rooms where texts were stored. The ink has faded in places, but the shape of the writing can still be followed.

Today, these fragments are kept in archives where the air is dry and the light is low. Curators turn them slowly beneath glass, studying the faint lines left behind by careful hands. The records are small and incomplete, yet they carry the quiet trace of someone who once walked those shelves, noting what stood in each place and what had been moved.

Chapter 8
The Doors That Closed at Dusk — Restricted Rooms and Quiet Limits

The last light of day fades along the high walls as attendants move through the halls, carrying small lamps from room to room. One by one, doors are pulled shut, their wooden frames meeting the stone with a dull, steady sound. A guard walks slowly down the corridor, checking each latch before stepping aside to let a librarian pass. Inside one room, shelves stretch from floor to ceiling, the air still and close, the smell of papyrus and dust heavy in the quiet.

A single lamp burns low as a bundle of scrolls is set carefully onto a table. The attendant steps back and waits while a key is turned in the door behind him. Outside, the noise of the city softens as night settles in. Within the closed rooms, the collection rests in silence, held behind doors that open only at certain hours.

Quiet Rules Within the Inner Rooms

As the collection grew larger, certain parts of the library were set aside for materials that were handled less often. Rooms near the heart of the complex were built with thick walls and narrow entrances, creating spaces where scrolls could be stored in steady conditions. Attendants moved through these corridors during the day, carrying bundles in and out while keeping careful note of where each text was placed. The movement of people was shaped by routine, with work beginning early and slowing as the light faded.

By evening, the inner rooms became quieter as fewer workers remained inside. Guards and trusted attendants checked the doors and watched over the halls as part of their daily duty. The library stood within the larger grounds of the Mouseion, close to the royal quarter, where officials kept a watchful presence. These rooms, set apart from the busiest areas, became part of the steady rhythm of work and care.

Scrolls Set Aside in Shadowed Rooms

Bundles placed in the inner rooms were made from papyrus sheets pressed and dried along the Nile, their surfaces smoothed for writing by careful hands. Some scrolls had been copied many times, the ink layered over earlier work, while others still carried the dark, fresh lines of recent writing. Scribes prepared each sheet with steady effort, cutting and shaping the material before rolling the finished text and tying it with cord. These texts were gathered and placed on shelves set deep within the rooms.

Attendants carried the scrolls along narrow corridors, lifting them with both hands and setting them down in spaces chosen for their safety. Tags were tied to the cords so they could be matched to written lists kept by the librarians. The rooms held older writings alongside newer copies, their surfaces marked by the work of many hands over many years.

Quiet Questions in the Corridor

Attendants moving through the halls began to notice which bundles were carried into the inner rooms and placed on the higher shelves. Some paused as the doors were opened, catching a glimpse of tightly packed scrolls set apart from the rest. A few spoke in low voices about why certain texts were stored in these spaces while others remained in the open rooms. The quiet handling drew attention, even as the work continued without explanation.

Among the scholars, small moments of hesitation appeared when requests were made to see materials kept behind those doors. Some waited outside while attendants stepped inside to check the shelves. Others exchanged brief looks, unsure how often the rooms were opened or who decided what was placed there. The uncertainty moved through the halls in careful glances and soft conversations.

Uneasy Watch Over the Inner Doors

As more texts were moved into the inner rooms, the presence of guards at the entrances became more noticeable. Officials passed through the corridors more often, stopping to check the locks and speak quietly with the attendants on duty. The growing need to watch over these spaces brought a sense of weight to the work. Those who moved through the halls could feel the shift as certain doors were opened only at set times and under careful watch.

Scholars began to notice the changes as well. Some waited longer than before to receive the scrolls they had asked to see, standing in the corridor while attendants disappeared behind closed doors. A few exchanged worried looks, wondering why access seemed tighter. In hushed voices, they spoke of how the handling of certain texts had become more guarded, and the feeling of pressure lingered in the narrow passageways.

Requests Made at the Locked Door

Records from the library describe how scholars would submit written requests when they wished to see texts stored in the inner rooms. Attendants checked these notes before unlocking the doors and carrying the scrolls out to reading tables. The process was slow and careful, shaped by routine and close attention. Some texts were brought out only for short periods, then returned to the shelves by the same hands that had carried them in.

Later writers claimed that certain works were kept hidden or restricted because of what they contained, though these stories were written long after the early years and cannot always be confirmed. What can be shown in earlier records is the steady use of written requests, the presence of guards, and the careful movement of scrolls in and out of the rooms.

Waiting in the Hallway's Silence

For the attendants who worked the inner doors, the responsibility settled heavily over time. Each day they checked requests, unlocked rooms, and carried scrolls back and forth with careful hands. Their movements were watched by scholars waiting in the corridor, and any delay brought quiet glances or soft questions. Some attendants felt the strain of being the ones who decided when a door opened and when it stayed closed, knowing their actions would be noticed.

Among the scholars, waiting became part of the routine. Some stood alone near the locked rooms, holding their tablets while they waited for a scroll to be brought out. A few felt the weight of being turned away when a request could not be met at once. Conversations grew shorter, and some chose to work apart from others rather than wait in crowded halls. The quiet pressure followed them through the day.

Routines Written Into Lock and Key

Over time, the handling of the inner rooms followed clear steps passed down through written instructions. Guards were assigned to certain doors, and attendants carried lists showing which texts were kept behind them. Requests were checked before a key was brought out, and entries were made each time a scroll was taken from the shelves. The movement of these materials followed a set path, from storage to reading tables and back again.

Some bundles were shifted deeper into the rooms as the collection changed, and a few doors were opened less often than others. The routines remained steady, shaped by written notes and quiet oversight. The process of carrying, recording, and returning the texts continued day after day, guided by the same careful steps.

Rooms Remembered by Those Who Waited

Years passed, and the practice of keeping certain texts in inner rooms remained part of the daily work. New attendants learned the same routines, carrying keys and checking written lists as they opened and closed the doors. Scholars who had once stood waiting in the corridors grew older and told others about the quiet spaces where some scrolls were kept. The habit of careful handling continued, passed from one group to the next as the halls changed with time.

Yet small questions lingered in the silence of those rooms. Some wondered how often the shelves had been moved or which bundles had been placed deeper inside over the years. A few entries in the records no longer matched what could be found, and no one could say when the change had begun. The sense of uncertainty stayed close, carried in the quiet sound of keys turning in the locks.

Doorways That Left Only Faint Signs

Archaeological work in Alexandria has uncovered fragments of walls and narrow corridors that suggest rooms once set apart from the larger halls. In some areas, the stone floors show worn paths where people walked the same routes over many years. Pieces of broken shelving and small tags made from thin materials have been found in nearby layers of earth, hinting at places where scrolls were once stored and handled with care.

In modern collections, curators study fragments of papyrus that carry marks of storage and handling, such as creases from being tightly rolled or faint impressions from cords that once held them closed. These pieces are kept in controlled rooms, where the air is dry and the light is soft. Even without the doors or keys that once guarded them, the traces of careful keeping still remain in the fragile surfaces that have endured.

Chapter 9
The Scroll That Vanished – A Text No One Could Find Again

A narrow beam of morning light falls across an open table where a single scroll lies unrolled, its edge curling slightly in the still air. A scribe leans over the surface, copying a line with careful strokes while another waits beside him, watching the ink dry. Outside the room, footsteps pass along the corridor, and the faint sound of voices drifts through the doorway. The work moves slowly, guided by quiet focus and steady hands.

Hours later, the table stands empty. The cord that once held the scroll together rests where it was set down, but the text itself is gone. Attendants move through the room, lifting bundles and checking nearby shelves. One opens a record tablet and traces a finger down a list, searching for a mark that might show where it was placed. The space feels strangely still as the search begins, the absence noticed in the hush that settles over the room.

A Hall Where Every Text Had a Place

Within the main rooms of the library, the work of copying and storing scrolls followed a steady pattern shaped by routine. Tables were set near the shelves so texts could be carried only a short distance when they were brought out for study. Attendants moved through the halls with quiet purpose, checking lists and returning bundles to their places. The rhythm of the work depended on careful record keeping and steady hands.

The collection had grown large enough that each room held rows of shelves marked by subject and origin. Scribes and scholars worked side by side, reading and writing under the light from high windows. Requests were carried on tablets, and attendants followed set paths to gather the scrolls that had been asked for. The movement of each text was noted as part of the daily routine.

A Scroll Formed by Many Hands

The missing text had been written on long sheets of papyrus pressed from reeds gathered along the Nile. The surface was smoothed and joined into a single roll, wide enough to hold lines of careful script. Ink was mixed from soot and water, then applied with a sharpened reed pen. The writing showed steady strokes and even spacing, the kind that came from a trained hand working with patience.

Some parts of the scroll had been copied from older writings, shaped through the work of more than one scribe over time. Small notes in the margins hinted that it had been read and handled more than once. The cord tied around it was worn where it had been opened and closed many times. It had passed through hands that carried it from table to shelf and back again.

A Quiet Realization at the Table

The first to notice the absence was the scribe who returned to the table where the scroll had been left. He paused, looking at the empty space where the papyrus had rested only hours before. His hand moved slowly over the surface as if expecting to feel the edge of the roll. He checked beneath nearby bundles and looked toward the shelf where it had been placed earlier, his movements growing more careful.

Attendants nearby began to take notice as he spoke in a low voice. One stepped forward to search the nearest rows, while another opened a record tablet to trace the last entry. A few scholars paused in their work, watching as the search spread from one shelf to another. There was no sign of damage or disorder, only a missing place where the scroll had once been.

Quiet Questions in the Corridors

As word of the missing scroll moved through the rooms, senior attendants began walking the shelves with more careful steps. They checked the lists and compared them with the bundles in each place, pausing at tables where the text had been read earlier.

A few scholars gathered near the doorway, speaking in low voices as they tried to recall who had handled the scroll last. The search spread slowly from one room to another, carried by careful glances and hushed conversation.

Soon the concern reached the offices where records were kept. Messengers were sent to ask for the lists to be brought forward, and those responsible for the shelves were called to answer quiet questions. The weight of the missing entry settled into the halls as more people joined the search. Eyes followed the attendants as they moved through the rooms, checking every space where the scroll might have been placed.

Lists Checked and Doors Watched

Attendants gathered the record tablets and began tracing the last known movements of the scroll. Each entry was read slowly, noting when the text had been brought to the table and which shelf it had come from. Workers moved from room to room, opening bundles and unrolling nearby scrolls just enough to check their first lines. Guards were asked to watch the doors as the search continued, and attendants followed the same paths again and again.

Later writers claimed that some believed the scroll had been taken on purpose, though these stories were recorded long after the event and cannot be confirmed. What is shown in earlier records is the steady effort to check the lists, question those who had handled the text, and look through the rooms where it had last been seen.

Eyes That Turned Toward One Another

The attendants who had worked near the table felt the weight of the search settle on them. Some moved more slowly through the rooms, aware that others were watching as they checked the shelves again. A few avoided speaking unless they were asked a direct question, keeping their focus on the tasks in front of them. The missing scroll left a quiet tension in the air, and the usual rhythm of the work seemed to falter as eyes followed each careful movement.

Among the scribes, there were moments of unease as they returned to their tables. Some worried that a small mistake might have led to the loss, and they looked over their notes more than once. Conversations grew short, and a few chose to work apart from the others. The absence of the scroll lingered in the room, carried in the silence that followed each careful search.

New Entries Written Into the Records

In the days that followed, the offices responsible for the collection began updating the written lists with greater care. Attendants were asked to note the movement of each scroll more closely, marking when it was taken from a shelf and when it was returned. Requests to handle certain texts were written down and checked before they were carried out. Tablets filled with fresh entries were carried between rooms as part of the effort to keep the records clear.

Some areas were watched more closely, and attendants were directed to follow set paths when bringing texts to the tables. Doors were checked more often, and the lists were read aloud to confirm what had been handled. The routines of recording and checking continued day after day, shaped by written instructions passed quietly from one office to another.

A Gap That Lingered in the Lists

Years later, the records still carried a small space where the missing scroll had once been marked. New attendants learned to follow the lists as they stood, yet a few older entries no longer matched what could be found on the shelves. Some spoke of the loss in quiet moments, recalling the day when the search moved through every room. The routines of checking and recording continued, shaped by the memory of that absence.

Even decades after the event, there were times when scholars paused over the records, tracing the lines where the scroll had been noted. No clear answer had been written beside the empty space, only the steady marks of later entries added around it. The sense of uncertainty stayed close, carried forward through the careful work of those who followed.

Faint Marks Left in the Records

Only small signs remain to show where a text once stood among the shelves. Fragments of old record lists, preserved in later copies and scattered references, still carry short entries that note titles now unknown. Some of these pieces show spaces where a name was written and later marked over or left unfinished. The thin sheets, worn by time, hold traces of hands that tried to keep the order of the collection clear.

In modern archives, these fragments are kept in quiet rooms where they are handled with steady care. Scholars study the faded lines under soft light, following the marks that once guided the movement of texts through the halls. The missing names, written and then lost, remain part of the story carried by the fragile pieces that survived.

Chapter 10
The Emperor's Visit – A Changing World Around the Library

Sunlight spills across the wide courtyard as soldiers take their places along the edges of the royal quarter. Their armor glints in the morning light, and the sound of their steps carries through the streets as they move into position. Attendants hurry along the paths, clearing the way and speaking in low voices as they pass. Inside the halls, the steady work slows as people pause, listening to the distant noise growing closer.

From the upper windows, figures watch as a procession moves through the city. Banners lift in the breeze, and the rhythm of marching feet echoes between the buildings. Near the entrance to the library complex, doors stand open as officials gather, waiting in stillness. Within the rooms beyond, scribes set aside their pens and glance toward the light, aware that something larger is passing through the city around them.

A City Under New Authority

In the early years of Roman rule, Alexandria stood as a busy and restless city shaped by change. Officials sent from Rome moved through the streets near the royal quarter, their presence marking a new order in daily life. Soldiers stood at key entrances, watching the flow of people and goods. Within the grounds of the Mouseion, the work of reading, copying, and storing texts continued, carried out by scholars who had long lived in its halls.

The library remained part of this larger setting, close to the center of power where decisions were made. Attendants carried scrolls between rooms, following the same paths they had walked for years. Messengers passed through the gates with written notes, and voices echoed in the corridors as routines went on. Outside, the city shifted around them, but inside the halls the work of keeping the collection in order continued.

Scrolls Shaped in a Changing Time

In the workrooms of the library, scribes continued preparing new texts on sheets of papyrus cut from reeds gathered along the Nile. The material was pressed and dried, then smoothed so ink would hold its lines. Each scroll was formed from joined sheets, rolled tight around a wooden rod, and tied with cord to keep it secure. Some were copies of older writings brought from distant lands, while others were newly written works shaped by the scholars who lived in the city.

Ink was mixed fresh each day, and the reed pens were trimmed to keep their points sharp. Scribes worked at long tables under steady light, reading from older scrolls as they copied the words onto new surfaces. Attendants carried finished texts away once the ink had dried, placing them among the growing collection. The work of shaping these scrolls moved forward through steady hands and patient labor.

Eyes Turning Toward the Procession

Scholars standing near the windows were among the first to notice the movement in the streets below. They watched as soldiers and officials passed through the city, their steps steady and their voices low. A few scribes paused in their work, glancing toward the courtyard where attendants had gathered. The sound of boots on stone carried into the halls, drawing quiet attention from those who worked among the shelves.

In the rooms beyond, attendants spoke in hushed tones as they carried scrolls from one table to another. Some wondered how the visit might affect their routines, unsure if new orders would follow. A few stood still for a moment, listening as the noise moved closer, then returned to their tasks with careful hands. A sense of waiting settled into the air as the procession neared the royal quarter.

Quiet Tension in the Courtyard

As the procession drew nearer, senior attendants moved more quickly through the halls, passing short instructions from one room to another. Doors were checked, and the paths between the shelves were cleared of carts and bundles. A few officials stood near the entrances, speaking in low voices as they watched the movement outside. The presence of soldiers close to the royal quarter brought a quiet weight to the air, felt by those who worked within the library.

Among the scholars, conversations grew softer as the sound of marching carried through the courtyard. Some set aside their pens and waited, unsure if their work might be paused or changed. A few gathered near the inner rooms, watching the doors as attendants moved past them. The steady rhythm of the day began to shift as the visit approached, and a sense of pressure settled over the rooms.

Lists Reviewed Under Watchful Eyes

As the visit moved through the royal quarter, attendants gathered record tablets and brought them to the main rooms where senior officials could see them. The lists showing what texts were held and where they were placed were read slowly and checked against the shelves. Workers moved from table to table, opening bundles and confirming entries as part of the routine. Records from the time describe the steady use of these lists to keep track of the collection.

Later writers claimed that outside authorities took a closer interest in certain writings during this period, though these reports were recorded long after the events and cannot always be confirmed. What can be shown is that the records were handled with care, and the movement of scrolls was watched closely as the city changed around the library.

Faces That Learned to Watch and Wait

As officials moved through the halls, the daily work took on a careful stillness. Scribes sat straighter at their tables, their hands steady but slower as they wrote. Some paused before unrolling a scroll, listening for footsteps in the corridor. The sense of being watched settled quietly into the room, and the usual soft conversations faded into silence. A few workers chose to keep to themselves, speaking only when needed, their focus fixed on the lines of ink before them.

Attendants who carried bundles from one place to another felt the strain of every movement. They checked the cords twice before lifting a scroll and walked more carefully through the narrow paths. A few worried that a mistake might draw attention at the wrong moment. The pressure followed them as they moved through the halls, carried in the quiet glances they exchanged as they passed.

Written Orders Passed Between Offices

In the weeks that followed the visit, new instructions moved quietly through the offices connected to the library. Messengers carried tablets from one room to another, and attendants were told to follow set steps when handling scrolls. Lists were checked more often, and the movement of texts was written down with greater care. Doors to certain rooms were opened only when needed, and entries were made each time a bundle was carried in or out.

Some texts were moved from one part of the building to another as space was adjusted and shelves were cleared. Attendants followed the written notes that guided where each scroll should be placed. The routines of carrying, recording, and returning the texts continued, shaped by instructions passed from one office to the next.

Changes That Stayed in the Records

In the years that followed, the routines shaped during that visit remained part of the daily work. New attendants were trained to check the lists more carefully and to note each movement of a scroll as it passed from shelf to table and back again. The habits settled into place and were carried forward as the collection continued to change. The halls returned to their steady rhythm, but the careful steps remained.

Even decades later, there were moments when the older records showed signs of those changes. Some entries were written more tightly, and a few shelves held bundles placed deeper into the rooms than before. Workers noticed the shift but could not always say when it had begun. The sense of quiet watchfulness stayed close, carried in the way the records were kept and read.

Faint Lines Preserved in Storage

Today, only scattered fragments remain that hint at how records were kept during those years of change. Pieces of papyrus found in later excavations still carry lines of careful writing that once tracked the movement of texts. Some fragments show repeated markings, as if lists had been checked again and again. The ink has faded in many places, but the shapes of the letters remain, pressed into the thin surface by steady hands.

In archives and museums, these pieces are stored in quiet rooms where the air is kept dry and the light is low. Curators lift them with tools made for delicate work, placing them between layers of protective glass. The records they hold are small and incomplete, yet they still carry a trace of the routines that once shaped the care of the collection.

Chapter 11
The Temple Without Scrolls – The Serapeum and the Last Collections

Under the bright afternoon sun, the steps of the Serapeum rise high above the city, its wide stone terraces warm to the touch. People move slowly along the pathways, some pausing in the shade of tall columns that stretch toward the sky. Inside, the air is cooler. Footsteps echo through long corridors where attendants pass quietly, carrying scrolls held close against their robes. Rooms open one after another, their shelves lined with the remains of collections gathered years earlier. Dust settles on the edges of wooden cases while hands work carefully to shift and store what still survives.

Beyond the walls, the streets of Alexandria are changing. Voices carry arguments about faith and authority, and new leaders issue orders that reshape public life. Inside the Serapeum, the work continues in careful silence. A scribe smooths a fragile sheet of papyrus across a table, checking its edges before placing it beside others already worn with age. Each movement is steady, measured, and watchful, as if the people within know that the safety of these rooms cannot be taken for granted.

A Sanctuary Above the City

High on the ridge overlooking Alexandria, the Serapeum stood apart from the crowded streets below. Its stone walls enclosed courtyards, chambers, and storerooms that had grown over time, shaped by worship and learning alike. Pilgrims climbed the long stairways to offer prayers, while attendants moved quietly through inner rooms, tending lamps and maintaining order. Within the complex, certain spaces were set aside where scrolls from earlier generations were kept, protected from the damp air near the harbor.

For many years, the Serapeum served both as a sacred place and a place of study. Teachers met in small groups beneath shaded porticoes, and scribes carried texts between rooms, recording and arranging them as carefully as space allowed. The building stood as a reminder of older traditions, supported by those who still valued its role. Even as the city's leadership shifted and new customs spread through the streets, the Serapeum remained a steady presence above the city, its halls still tied to learning as well as devotion.

Pages Carried from the Earlier Halls

Many of the scrolls kept within the Serapeum had been written years before, copied by trained hands in rooms once filled with steady light and careful movement. Scribes worked at long tables, smoothing sheets of papyrus and trimming rough edges so the pages would lie flat. Ink was mixed from soot and plant resin, thick enough to leave dark, clear lines that could last for years if kept dry. Each text was formed piece by piece, joined into long rolls and tied with cord before being placed into storage.

Assistants carried these finished scrolls from one building to another as the needs of the collection changed. Some were moved from older rooms that could no longer hold them safely, while others were brought in from smaller study spaces across the city. Hands passed them carefully along corridors, lifting them onto shelves within the Serapeum's inner chambers, where they could be kept away from weather, damage, and the steady wear of daily use.

Whispers Among the Shelves

Word spread quietly among those who worked in the Serapeum as certain scrolls were brought into its inner rooms. Attendants noticed the change first, watching as bundles from older collections were carried in and set aside for careful storage. Some paused as they passed, glancing at the worn bindings and faded labels that marked them as older than most. A few spoke in low voices, wondering how long these texts would remain safe here, away from the busier halls below.

Teachers and scribes began to pay closer attention as they requested access to the stored writings. Some handled them with extra care, as if the act of unrolling them might cause further damage. Others hesitated before taking them out at all, uncertain how often they should be used. Quiet concern settled in as the flow of new texts slowed, and the movement of older scrolls into the Serapeum felt less like expansion and more like sheltering what remained.

Tension in the Courtyards

In the streets below the hill, arguments grew louder as the city's religious life shifted. Groups gathered in public squares, speaking openly about which customs should remain and which should be set aside. The Serapeum, long tied to older forms of worship, began to draw attention. Officials and local leaders passed by its steps more often, pausing to look toward its columns as if measuring its place in the changing order of the city.

Inside, those who worked among the shelves sensed the shift in tone. Visitors arrived with more questions than before, asking about the building's purpose and what was kept within its rooms. Some teachers spoke in hushed voices about the growing pressure to step away from older traditions. Others moved carefully through their work, uncertain how long the protection that once surrounded the Serapeum would continue. The sense of watchfulness deepened as days passed and the voices outside grew harder to ignore.

Hands That Guarded the Written Words

Within the Serapeum, access to certain scrolls began to change. Attendants kept closer watch over the rooms where older texts were stored, noting who entered and what was removed. Teachers requested permission before unrolling rare works, and scribes recorded which items were taken out and returned. The careful tracking of movement was meant to prevent loss or damage, and some accounts from the time mention that texts were placed in inner chambers where fewer people could reach them.

Later writers claimed that certain works were hidden or sealed away to keep them from harm, but clear records of such actions are limited. What is known comes from scattered notes and references that describe the tightening of access as tensions grew in the city. Some teachers spoke of being asked to return scrolls more quickly than before, and a few reported that certain rooms were no longer open for study at all, suggesting quiet efforts to control how the texts were handled.

Lives Marked by the Work They Chose

Those who spent their days inside the Serapeum began to feel the strain beyond its walls. Some attendants noticed neighbors speaking less to them, and a few found that old friendships grew distant.

Their work had once been respected, but now it set them apart. Teachers who continued to request access to stored texts sensed the change in tone when they walked through the streets. Conversations fell quiet as they passed, and certain invitations stopped coming.

A few scribes chose to visit the building at different hours to avoid attention, slipping through the gates when fewer people were watching. Others kept their roles private, careful not to speak openly about the rooms they tended or the scrolls they handled. Even within the Serapeum, tension showed in small ways. Workers spoke less freely, unsure who might repeat their words. The daily routine remained the same, but the weight of quiet judgment followed many of them home.

Orders Passed Through Quiet Channels

Over time, changes in handling became more formal. Attendants were given clear instructions about who could enter certain rooms and when. Lists were checked at doorways, and only those with specific tasks were allowed through. Some scrolls were moved from open shelves to inner storage spaces, where fewer people could reach them. A few rooms were closed entirely, their doors kept shut except for brief visits by those assigned to manage the contents.

Records mention that requests for access were answered more slowly than before. Teachers who once borrowed texts freely now waited for approval, and some were told to work only with copies already in their possession. Courtyard meetings became less common as gatherings were quietly reduced. Messages passed between officials and attendants without public notice, and decisions were carried out with little explanation. The routines of storage and handling continued, but the flow of movement through the building grew more controlled with each passing season.

Years of Absence and Quiet Memory

In the years that followed, people spoke of the Serapeum with a different tone. Some who had once worked there moved on to other roles, carrying with them memories of the rooms and shelves they had tended. Teachers who remained in the city found fewer places where older texts could be handled openly. A few continued to teach from copies they had kept, passing lessons along in smaller settings where attention was less likely to gather.

Decades later, uncertainty remained about what had been taken, what had been lost, and what might still have survived in private hands. Some recalled seeing scrolls carried out before the rooms were closed, while others spoke only of empty spaces where shelves had once stood. No single account offered a clear answer. The absence lingered in small ways, in the missing texts people once expected to find and in the quiet question of how much had disappeared without record.

Stones, Fragments, and Faded Records

Today, the ground where the Serapeum once stood still holds traces of its past. Broken columns, worn steps, and scattered foundation stones remain in place, marking where rooms once opened into one another. Excavations have uncovered parts of the structure, along with pieces of storage niches cut into stone walls. These spaces, once used to hold scrolls, now stand empty. Small objects found nearby, including fragments of pottery and tools, suggest the steady work that once filled the halls.

Written references preserved in later manuscripts offer brief mentions of the building and the collections it sheltered. These records are held in archives and studied carefully, their details compared to the physical remains found on the hill. The shelves are gone, and the scrolls once kept there cannot be traced with certainty, yet the stone chambers remain. Standing among the ruins, the quiet shape of the structure still suggests the careful movement of hands that once carried knowledge into its shadowed rooms.

Chapter 12
The Books That Scattered – Knowledge Without a Home

In a narrow room lit by a single window, a teacher sits at a low wooden table as a small group gathers close around him. Outside, the sound of carts rolling over stone streets rises and fades, but inside the room the only noise is the soft rustle of papyrus. A worn scroll is unrolled with care, its edges darkened by years of use. The teacher pauses often, reading slowly so each word can be followed. The room is small, and the shelves hold only a handful of texts, each one kept within reach.

Light shifts across the floor as the afternoon passes. A student leans forward to see the lines more clearly, while another copies a passage onto a fresh sheet. These writings have moved from place to place, carried in bundles through crowded streets and passed from one set of hands to another. No large hall shelters them now. Each scroll rests where it can be watched, used, and protected, held close by those who know it may not be replaced if it disappears.

Learning in Small Rooms Across the City

After the great centers of collection fell quiet, teaching did not stop in Alexandria. Schools continued in modest spaces near markets and along side streets where students gathered in small numbers. Some lessons took place in private homes, others in open courtyards where benches and tables were set close together. Without large halls to hold thousands of scrolls, instruction relied on what could be kept nearby and handled often.

Teachers arranged their work around the few texts they could store safely. Shelves were built into walls, and chests were placed beneath tables to protect fragile rolls from dust and damp air. A handful of students might gather at a time, each waiting for a chance to read from the same worn copy. These smaller places formed a loose network of learning across the city, shaped by the effort of individuals who kept study alive without the support of a single central institution.

Copies Made by Hand and Passed Along

Many of the texts used in these small schools had been copied years earlier, often by students learning the craft of writing. Sheets of papyrus were cut and joined into long strips, then rolled tightly and tied with cord to keep them from fraying. Ink was mixed from soot and water, thick enough to leave dark marks that would not fade quickly. Each line was written slowly, with care taken to keep the words straight and clear.

The work of copying was steady and demanding. One person read aloud while another followed the sound of each word, setting it down on a fresh surface. Mistakes were crossed out or scraped away, then written again. Some texts were copied from older versions that had already passed through many hands, their letters worn but still readable. Over time, these repeated efforts created new scrolls that carried forward the same words, shaped by patience, time, and the skill of those who made them.

First Glances at the Worn Pages

Students were often the first to notice how different these texts looked from the ones their teachers described from earlier years. The edges were rough, the ink uneven in places, and some lines faded where hands had touched them again and again. As the scrolls were unrolled in small rooms, learners leaned closer, studying each mark with care. A few asked where the copies had come from, their voices low, unsure if the question was welcome.

Teachers watched closely as the scrolls passed from one set of hands to another. Some paused before allowing a student to hold one, as if weighing the risk. There was quiet concern in the way the texts were handled, each crease and tear noticed more than before. A few teachers spoke softly among themselves about how few complete works remained. Their voices carried a sense of hesitation, as though they were still adjusting to the idea that what lay before them might be all that was left.

Uneasy Watch Over the Lessons

As teaching continued in small rooms across the city, some neighbors began to take notice of the gatherings. Groups of students arriving at the same doors day after day drew quiet attention. A few people paused nearby, listening to the sound of voices reading aloud from older writings. Questions were asked about what was being taught and which texts were being used. In some places, owners of buildings grew cautious, uncertain how much notice these lessons might attract.

Among teachers, there was a growing sense of strain. A few spoke of visitors who lingered at the edge of a lesson, watching without entering. Others mentioned being asked to explain the purpose of their work or the origin of the scrolls they used. Some chose to keep their classes smaller and less visible, spacing them out across different hours. Conversations between peers carried a tone of concern, as if the simple act of reading from certain texts now required careful thought and quiet restraint.

Hands That Copied and Hands That Checked

Teachers and scribes worked to study the texts with care, often comparing one copy against another to catch errors that had crept in over time. A reader would speak each line aloud while a second person followed the written words, pausing when something sounded different. Marks were added in the margins, and missing lines were filled in when another copy could be found. These efforts were meant to keep the wording steady, even as each new version carried small changes.

Some later writers claimed that certain lessons were watched or restricted by local officials, but clear records of strict control are limited. What survives in letters and notes shows that some teachers chose to keep certain texts within trusted circles, sharing them only with students they knew well. In a few cases, texts were copied more than once so that they would not be lost if one version wore out. Access was shaped by caution, habit, and the quiet choices of those who handled them.

The Quiet Weight of Staying

Teachers who continued their lessons began to feel the strain in small, personal ways. Some found that fewer students came to their doors, as families grew cautious about where they sent their children. A few instructors heard whispers in the streets, suggesting that their work belonged to an older time. Invitations to speak or teach in public spaces became less frequent, and the circle of those who supported them slowly narrowed.

Students felt the change as well. Some were warned at home to be careful about what they studied and where they went each day. A few stopped attending altogether, leaving empty spaces at the table where they once sat. Those who remained grew quieter, speaking less about their lessons outside the room. Even within the small schools, there was a sense of distance forming, as if each person carried a private concern about how long they could continue before attention turned too closely in their direction.

Quiet Rules and Careful Handling

Over time, those responsible for maintaining order in different districts began to set quiet limits on how and where teaching could take place. Some gatherings were asked to move indoors, away from open courtyards where voices carried into the street. A few property owners required notice before rooms could be used for regular lessons. These changes were passed along in simple terms, often without written orders, and were followed to avoid drawing further attention.

Teachers adjusted their routines to fit these new conditions. Scrolls were stored in chests that could be closed and carried if needed. Some texts were moved between homes to prevent them from remaining in one place for too long. Requests to borrow certain writings were handled more carefully, and records of who held them were kept in small personal lists. The work of copying and teaching continued, shaped by a growing pattern of quiet management and cautious handling.

Years That Passed Without a Central Place

As the years moved on, those who had once learned in the small rooms grew older and carried what they remembered into new places. Some became teachers themselves, using the same worn copies they had studied from as younger students. The lessons spread slowly from one neighborhood to another, shaped by memory and the care taken to keep the words alive. Without a single building to hold everything together, learning settled into scattered pockets across the city.

Decades later, people still spoke of how much had once been gathered in one place, though few could say exactly what had been lost. Some texts that had been copied many times began to show gaps where lines were missing or hard to read. Teachers wondered if the words had changed over time, or if something important had slipped away without notice. The uncertainty lingered, carried in quiet questions that had no clear answers.

Fragments That Outlived Their Rooms

In modern collections, a few surviving papyrus fragments show signs of long use. Their edges are worn thin where fingers once held them, and faint repairs can be seen where small tears were joined with paste. Some of these pieces were found far from the city where they were first written, preserved in dry ground or stored in later archives. The handwriting varies from one copy to another, suggesting that many hands worked over time to keep the words readable.

Archivists now keep these fragments in controlled rooms, laid flat and protected from light and air. Notes beside them record where they were found and how they were handled in earlier centuries. Each mark and crease offers a quiet trace of the people who copied, carried, and studied them. Though the rooms where they were first read are gone, the lines remain, holding the shape of voices that once gathered close to listen.

Chapter 13
The Names That Endured – Scholars Who Carried the Knowledge

In a quiet study set along a narrow street, a scholar bends over a wooden table covered with scrolls and loose sheets of papyrus. Light from a small oil lamp flickers across the surface, casting moving shadows over lines of careful writing. Outside, footsteps pass in the dark, but inside the room the only sound is the soft scratch of a reed pen. A stack of copied pages rests at one side, each marked with notes written in the margins by hands from earlier years.

He pauses often, lifting one scroll and then another, comparing words and checking passages before copying them again. These texts have traveled a long path to reach this table, carried through cities and across generations. Some bear names of earlier teachers written at the top, reminders of who once studied and preserved them. The room is small, the shelves few, yet each page holds the work of many lives. Hands move steadily, as if the act of writing is all that keeps the words from slipping away.

Learning Carried by Individuals

Across the city and beyond its walls, study continued without the support of large halls or royal funding. Teachers worked from small rooms, and scholars often lived close to where they taught. Some gathered students in courtyards, while others opened their homes to those willing to learn. Each place held only a few scrolls, stored in chests or stacked on narrow shelves. The work moved from one person to another, shaped by memory, effort, and the quiet habits of daily study.

In this setting, the role of the scholar changed. Instead of working within a single institution, many relied on personal connections to keep learning alive. They traveled between cities, visiting friends and former students, carrying notes and copies with them. A name written at the top of a page became a sign of trust and recognition. These scattered efforts formed a loose network that kept knowledge moving, even as the larger centers that once supported study no longer stood at the heart of the work.

Pages Written by Careful Hands

Many of the texts used by these scholars were created through steady, repeated copying. Papyrus sheets were cut into long strips and pressed flat before being joined into rolls. A scribe mixed ink from soot and water, testing its thickness on a scrap piece before beginning. Lines were written slowly from left to right, each word formed with care so the letters would remain clear for future readers. When one roll was finished, it was dried, smoothed, and tied to keep its shape.

Some texts were copied from older versions that had already been handled for years. Edges were trimmed where they had worn thin, and weak places were reinforced with small patches. A reader often spoke the words aloud while another wrote, keeping the pace steady so lines would not be skipped. Over time, new copies formed from these efforts, shaped by patience and long hours at the table. Each scroll carried the marks of the hands that made it and the voices that guided its creation.

First Readers and Quiet Questions

Students were often the first to notice when a new copy appeared. They gathered close as the scroll was unrolled, studying the fresh ink and the careful spacing of the lines. Some traced the letters lightly with their eyes, recognizing passages they had heard spoken before. Others paused, unsure if the wording matched what they remembered. The teacher watched closely, listening for hesitation as the first lines were read aloud.

Fellow scholars also took interest when a copy was shared. A few leaned over the table to compare it with older versions they carried, pointing to places where the words seemed slightly different. Conversations stayed low and careful. No one spoke with full certainty about which version was closest to the original. Some wondered if parts had been lost in the copying, while others worried about using a text that might not match what earlier teachers had known.

Watchful Eyes in Quiet Places

As copies of certain texts began to pass from one scholar to another, attention followed them. In some neighborhoods, gatherings of students drew notice from those who passed by. People paused at doorways, listening to the sound of voices reading aloud from inside. A few asked questions about what was being taught and where the scrolls had come from. These moments were brief, but they left a sense of being watched.

Among scholars, concern grew in small, careful ways. Some spoke in low tones about choosing their words more carefully during lessons. Others became cautious about who they allowed to see their copies, unsure how their work might be received beyond their own circles. Even among peers, conversations shifted. A scholar might hesitate before sharing a new copy, wondering if the text might draw unwanted attention. The act of teaching began to carry a quiet weight that had not been there before.

Careful Reading and Guarded Use

Scholars worked with the texts in a measured way, reading passages aloud and checking them against other copies when they could. Some kept small notebooks beside them, marking places where words seemed unclear or where lines differed from what they remembered. These notes were kept close and shared only with a few trusted students. In certain rooms, teachers chose which sections could be read openly and which were set aside for private study.

Later accounts sometimes claimed that specific works were restricted by local leaders, but clear records are scarce. What survives in letters and scattered references suggests that some scholars limited access on their own, deciding who could see a text and when it would be used. A few copies were wrapped and stored carefully between lessons, taken out only when needed. The handling of these writings grew more deliberate, shaped by caution and the steady effort to keep them from being lost or damaged.

The Weight of Being Known for the Work

Scholars who spent their time copying and teaching older texts began to feel the strain in their daily lives. Some noticed that neighbors grew distant, speaking less freely when they passed in the street. A few were no longer asked to join public gatherings, and invitations that once came easily stopped arriving. Their work had become something that set them apart, and the quiet attention that followed them carried an uneasy weight.

Among friends and family, tension sometimes appeared in small, private moments. A relative might ask why the work continued when it brought so little support. Students who stayed close to these teachers felt the change as well, aware that their lessons were not always welcomed by others. Even within their own circles, scholars spoke carefully, unsure how their words might be repeated. The simple act of carrying a scroll through the street could draw a glance that lingered longer than before.

Quiet Measures and Changing Routines

Over time, some local authorities began to pay closer attention to gatherings where teaching took place. In certain districts, meetings were asked to remain small and to avoid public courtyards where voices could carry. Property owners were reminded to keep track of who used their rooms for lessons. These changes were often passed along by word of mouth rather than written orders, but they shaped how and where scholars worked.

Within this setting, teachers adjusted how they handled their texts. Scrolls were moved between homes so they would not remain in one place for too long. Some were stored in chests that could be closed and hidden quickly if needed. Lists were kept of who borrowed copies and when they were returned. A few lessons were shifted to different hours to avoid notice. The routines grew more careful, shaped by quiet rules and the need to manage each text with steady attention.

Years That Carried Their Names Forward

In the years that followed, the names of certain scholars remained tied to the texts they had copied and taught. Students who had learned under them carried those names into new cities, repeating them as they shared the writings with others. Some copies bore small notes marking who had written them or who had passed them along. These marks helped later readers trace the path of a text, even when the place it first came from was no longer certain.

Decades later, questions remained about how much had been changed as the words moved from hand to hand. Some passages appeared slightly different in one copy than in another, and no one could say for sure which version matched the earliest writing. Teachers spoke quietly about these differences, unsure if they came from simple mistakes or from lines that had once been lost. The uncertainty stayed with the texts, carried forward with each new copy that bore the name of someone who had worked to preserve it.

Names Written in the Margins

In modern archives, some surviving manuscripts still carry faint notes written along their edges. These markings record the names of teachers, scribes, or places where the text once passed through. The ink is lighter than the main writing, added later by careful hands that wanted to be remembered or to guide the next reader. In a few cases, the handwriting changes from one section to another, showing that the scroll was copied more than once across many years.

Archivists now study these details under steady light, noting each variation in letter shape and spacing. The small differences reveal how the texts traveled from person to person, shaped by the labor of those who kept them alive. Some copies rest in controlled rooms, protected from air and moisture, while others remain in private collections known only through records. On the surface of each page, the names remain, quiet traces of the people who once held the words and passed them forward.

Chapter 14
The Memory of the Lost Halls — Stories That Grew Over Time

In a dim room filled with scattered notes, a writer leans over a table covered in copied passages from older works. A small lamp burns beside him, its light shifting as the flame moves in the still air. He reads slowly, tracing each line with his finger before setting it aside and reaching for another. Some pages speak of crowded halls filled with scrolls, others of fires, conflict, and sudden loss. None of the accounts tell the same story in quite the same way.

Outside, the city hums with its usual rhythm, but inside the room the only sound is the soft turning of pages. The writer pauses often, comparing names, places, and events written years apart by different hands. Some details match closely. Others leave gaps that cannot be filled. Each fragment offers a glimpse of something once large and certain, now remembered in pieces. He gathers the notes carefully, arranging them into small stacks, working from what remains to shape a picture of halls he has never seen.

Rooms of Study After the Great Collection

In the years after the large halls were no longer in use, learning continued in scattered rooms across different cities. Writers worked in small studies, surrounded by shelves that held only a few texts at a time. Schools formed around teachers who gathered students close, reading from copies that had been carried from place to place. These settings were quieter and more limited, shaped by what could be stored and protected within reach.

Those who wrote about the old library lived in a world where the institution itself no longer stood at the center of daily life. They depended on earlier writings, stories passed down through teaching, and notes gathered from older sources. Some worked in private homes, others in small schools supported by local leaders. Each place held its own fragments of memory, and the effort to write about the past took place within these modest spaces where knowledge was kept alive through careful reading and steady work.

Pages Drawn from Older Accounts

Writers who set out to record the story of the lost halls worked from what they could gather. They copied lines from earlier scrolls, setting them down on fresh sheets of papyrus. Some used ink mixed in small bowls beside their tables, dipping their pens and writing slowly to keep the words clear. Notes were added along the margins, marking where a passage came from or where another version said something different. The work took time, shaped by careful hands and steady attention.

Bundles of copied pages grew on the table as each section was written out. Assistants sometimes helped by reading older texts aloud while the writer followed along, setting each line into place. When a page was finished, it was laid aside to dry before being joined to others in a long roll. These new texts were formed piece by piece, built from earlier records that had passed through many hands before reaching the room where they were written again.

First Readers and Careful Questions

When these newly compiled accounts began to circulate, they drew quiet attention from teachers and students who were eager to learn more about the lost halls. A few copies were read aloud in small gatherings, the listeners leaning closer as the names of earlier rulers and scholars were spoken. Some nodded as they recognized parts of the story they had heard before. Others paused at certain lines, uncertain if the details matched what they had been taught.

Among fellow writers, the reactions were measured. Some compared the new text to older notes they kept in their own collections, tracing differences in wording or dates. A few spoke in low voices about how much could truly be known, aware that the passages came from sources written at different times. There was interest, but also hesitation. Each new account brought with it the quiet question of how much had been remembered clearly and how much had shifted as the story passed through many hands.

Watchful Interest from Outside Circles

As these written accounts began to pass from one reader to another, they reached beyond the small groups where they were first shared. Some teachers heard that their notes were being discussed in places they did not expect. A few visitors arrived with questions about where the information had come from and who had written it down. The attention was quiet but steady, and it carried a tone that made some pause before sharing their copies again.

Among peers, a sense of caution began to grow. Writers spoke more carefully about what they chose to include and how widely they allowed their work to travel. Some hesitated before naming certain sources, unsure how their words might be received. Conversations that once took place openly at tables moved into private corners and smaller rooms. The simple act of writing about the past now seemed to draw notice, and that notice brought a growing weight of concern.

Comparing Lines and Guarding the Copies

Writers and teachers worked carefully to study the accounts as they spread. Some placed two versions side by side on the table, reading each line aloud to check where the wording matched and where it did not. Notes were added in the margins to mark uncertain names or events that appeared differently in older records. A few copied the same passage more than once, hoping to keep the details steady as the text moved from hand to hand.

Later claims suggest that certain passages were set aside or shared only within trusted groups, though clear records of strict control are limited. What survives in letters and copied notes shows that some writers chose to keep parts of their work close, allowing only a few students to read them. A few copies were wrapped and stored in chests between readings. These choices shaped how the text was handled, guided by caution and the steady effort to keep it from being altered or lost.

Quiet Distance Among Colleagues

Writers who gathered accounts about the lost halls began to notice small changes in how others treated them. Some found that conversations grew shorter when their work was mentioned. A few were no longer asked to join public readings or teaching circles where they had once been welcomed. Their focus on recording the past set them apart, and the steady attention their writings received brought an unease that followed them into daily life.

Within their own circles, strain showed in careful silences. Friends who once shared notes freely now hesitated before passing along older records. Some writers chose to keep their work private, unsure how their names might be tied to the accounts they were gathering. Even at home, the work could bring tension, as family members worried about the attention it drew. The task of preserving memory began to feel heavier, carried quietly in the spaces where trust had once come easily.

Quiet Custody and Careful Circulation

Over time, some local schools and study houses began to keep closer track of the written accounts that passed through their rooms. Lists were kept of which copies were held and who had borrowed them. In a few places, certain texts were stored in locked chests and brought out only when a trusted reader asked to see them. The movement of these writings became more ordered, shaped by simple routines meant to prevent them from being misplaced or damaged.

In some districts, teachers were asked to keep gatherings small and to avoid public readings that might draw large crowds. Copies were sometimes moved from one place to another, passed quietly between rooms so that no single location held too many. Notes were added to the front of some scrolls to mark where they had come from and who was responsible for them. These small acts of recordkeeping and careful handling guided how the texts were kept and shared over time.

Echoes That Outlived the Writers

Years after the first accounts were written, later readers still turned to them to piece together what had once stood in the city. Copies were made again and again, each one carrying the names of earlier writers along its edges. In some schools, teachers spoke of those who had gathered the first records, repeating their names as part of the lesson. The writings became a quiet link across time, passed from one generation to the next.

Decades later, uncertainty remained about how much of the story had been captured fully. Some passages seemed to shift in detail from one copy to another, and no one could say which version was closest to the earliest notes. A few names appeared in certain texts but not in others, leaving gaps that could not be easily filled. The effort to remember continued, shaped by the pieces that survived and the quiet question of what had slipped away between the lines.

Pages Preserved in Later Hands

In modern archives, some of the accounts gathered long ago survive as fragile copies, their surfaces marked by years of handling. The papyrus has darkened, and the edges show where repairs were made to keep the sheets from breaking apart. Careful writing along the margins records the names of earlier owners and the places where the texts were once kept. These details offer a quiet trace of how the stories moved through time.

Archivists now store these pages in controlled rooms, laid flat beneath soft light and protected from air and moisture. Each line is studied, compared with other copies, and recorded to preserve what remains. Though the original halls no longer stand, these scattered pages still carry the effort of those who tried to remember. The marks left by their hands remain visible, resting in silence beneath the careful watch of those who guard them today.

Chapter 15
The Traces Beneath the Sand – Searching for the Lost Buildings

In the heat of the afternoon, the sound of metal tools tapping against stone breaks the quiet along the edge of the modern city. Workers kneel close to the ground, brushing away layers of sand with careful hands. A faint line of worn blocks begins to appear, their surfaces smoothed by time and buried for centuries. Nearby, a man bends over a small notebook, sketching the shape of the exposed stones while measuring the distance between them. The harbor lies not far away, its waters moving slowly as they have for generations.

Dust rises in thin clouds as more earth is cleared, revealing corners of old walls and fragments of steps that once led somewhere unknown. A piece of carved stone is lifted gently and set aside. Voices stay low, and every movement is measured. The ground beneath the city holds its secrets close, and each uncovered surface suggests the outline of spaces that once stood above. The work continues in steady silence, guided by patience and the hope that the buried past will show itself piece by piece.

A City Built Over Its Own Past

Over the centuries, the ground of Alexandria changed as new streets, homes, and public buildings rose above the remains of older ones. Layers of stone and soil settled over what had once been open courtyards and halls. People walked across these places each day, often unaware of what lay beneath their feet. The city continued to grow, shaped by new rulers, new needs, and the steady rhythm of life along the coast.

As the modern city took form, interest grew in the earlier landscape hidden below. Records from travelers and earlier writers gave rough ideas of where important buildings once stood. Small teams began to walk the edges of known sites, studying old maps and noting the rise and fall of the land. Work moved slowly through areas where space allowed, guided by careful planning and quiet effort to understand how the earlier city had once been arranged beneath the present one.

Notes Drawn from Stone and Soil

As work began at different sites, careful records were created to capture each new discovery. Workers brushed away sand and lifted loose soil while others watched closely, ready to mark what appeared. Lines were drawn on paper to show the shape of walls and floors as they emerged. Measurements were taken with simple tools, and each stone was noted for its size and position. The process was slow, shaped by steady hands and repeated checking to make sure nothing was missed.

These written records grew page by page as the work continued. Sketches showed the outlines of steps, corners, and broken surfaces found beneath the ground. Notes were added beside each drawing to describe what had been uncovered and how deep it lay. Some pages were copied and shared so others could compare what was found at different sites. Each sheet became part of a growing collection of information built from the careful labor of those working close to the earth.

First Eyes on the Emerging Records

As the first notes and sketches began to circulate among small groups of researchers, they drew quiet attention. A few scholars gathered around rough drawings spread across tables, tracing the lines that showed where walls had been found. Some leaned closer, pointing at markings that suggested doorways or steps. Others watched in silence, unsure what the shapes might truly represent. The information felt important, but no one spoke with full certainty about what had been uncovered.

Visitors who saw the early records reacted with careful interest. Some asked if the stones could belong to the buildings long spoken about in older accounts.

Others hesitated, aware that the ground held many layers from different times. There was a quiet sense of concern about drawing conclusions too quickly. Each new page brought hope, but also the weight of not knowing whether the fragments matched what people had been searching for over so many years.

Questions Raised by the Work

As word spread about the digging and the notes being kept, more people began to take an interest in what was being uncovered. Local officials visited the sites to see the work for themselves, asking where the stones had been found and how the records were being handled. Some wanted to know who had given permission for the digging and how long it would continue. Their questions were steady and direct, shaping the pace of the work as teams paused to explain what they were doing.

Among fellow researchers, concern grew in quiet conversations. Some worried that the attention might bring limits on where they could dig or what they could record. Others spoke of the need to be careful when sharing their notes, unsure how the information might be received by those who controlled the land. The growing interest placed a weight on the work, and each new discovery drew more eyes than before.

Lines Measured and Marked

Teams worked carefully to study the notes and drawings as they grew in number. Measurements were checked again and again, and new sketches were made to show how each section of stone connected to the next. Some records were kept in locked cases to protect them from damage. Copies were made so that more than one group could review the findings. Workers returned to the same places to confirm what they had first written, adjusting their notes when the ground revealed more detail.

Later stories sometimes claimed that certain discoveries were hidden or held back, but clear records mostly show a steady effort to protect fragile findings. Some sites were closed off to prevent damage from curious visitors, and access was limited to those assigned to the work. Notes were shared in small circles, and some drawings were kept from public view until they could be checked. These steps shaped how the records were handled as the work moved forward.

The Strain of the Search

Those who spent long days at the digging sites felt the weight of their work follow them home. Some were away from their families for weeks at a time, living close to the ground they were studying. The heat, the dust, and the steady focus left them tired, and their hands showed the marks of constant labor. A few spoke of being questioned often about what they were finding, repeating the same careful answers again and again.

Among friends and neighbors, the work sometimes set them apart. Some people believed the search disturbed places that should be left alone, and quiet disapproval showed in small ways. Invitations became fewer, and conversations turned cautious when the topic came up. Even within their own teams, tension appeared as pressure grew to find clear results. The long hours and the weight of expectation settled into daily life, carried quietly by those who stayed at the edge of the earth and kept digging.

Permits, Records, and Careful Oversight

As the work continued, formal procedures began to shape how the digging was carried out. Written permits were required before teams could begin work in certain areas. Officials kept records of which sections of land were opened and what was found there. Objects uncovered from the ground were listed and stored under supervision, with notes made about where each piece had been located. Access to some sites was limited to specific hours, and barriers were set in place to keep others from entering.

Reports were gathered and sent to central offices where they were reviewed and filed. Some materials were transferred to secure storage rooms for protection, where they could be kept safe from weather and damage. Teams were asked to document their work in detail, noting each step and recording each change. These measures shaped how the search moved forward, guided by routine and careful handling at every stage.

Years Marked by Unanswered Spaces

Years later, the early records and drawings from those first digs were still used by new teams trying to understand the buried city. Fresh workers compared their findings with older notes, tracing the same lines and measurements to see what matched. Some places once marked for study were no longer open, covered again by new streets and buildings. The earlier work became a guide, but it also reminded them of how much ground could no longer be reached.

Decades passed, and questions remained about what had been missed or left behind. A few notes mentioned walls that were only partly uncovered before the work stopped, and no one knew if they led to larger rooms beneath. Some early sketches showed shapes that could not be found again. The ground held its silence, and the spaces that had once been touched by careful hands faded from view, leaving behind only lines on paper and the memory of what might still lie below.

Fragments Mapped and Preserved

Today, pieces uncovered from the early searches remain stored in controlled rooms, each marked with labels that record where they were found. Stone blocks, carved fragments, and sections of foundation are kept on shelves or in secure storage areas. Old notebooks filled with sketches and measurements are placed in archives, their pages protected from light and moisture. Each entry shows where workers once brushed away sand and uncovered what lay beneath.

Modern teams return to these records when planning new work, tracing the old drawings and comparing them to what can still be seen on the ground. Some marks made long ago still match the outlines of buried walls beneath newer structures. The fragments sit quietly in their cases, worn by time but still present. In the steady light of storage rooms, the stone and paper remain, holding the shape of places that once stood above the earth.

Chapter 16
The Scrolls That Survived – Texts Preserved Beyond the City

In a dim archive far from the coast, a caretaker lifts the lid of a wooden chest and pauses before unrolling a fragile scroll. The surface is darkened with age, its edges worn where hands have held it many times. Light falls softly across the writing, revealing steady lines formed by careful work long ago. Shelves nearby hold other texts stacked side by side, each tied with cord and marked with faint labels. The air is still, carrying the dry scent of old paper and dust.

A small group gathers quietly as the scroll is opened further, leaning close to read what remains. Some lines are clear, others faded, but the words are still there. Outside, distant sounds of the city drift through narrow windows, yet inside the room the focus stays fixed on the page. These writings have traveled far, carried from place to place through years of change. Each mark on the surface speaks of the long path it has taken, passing through many hands before coming to rest here.

Quiet Rooms Far from the Harbor

In cities beyond Alexandria, learning found shelter in modest buildings set along narrow streets and quiet courtyards. Some were attached to schools where teachers gathered small groups of students. Others were private rooms kept by families who had long cared for written works. Shelves lined the walls, holding bundles of scrolls wrapped in cloth to keep them safe from dust and damp air. These spaces became places where knowledge was read, copied, and guarded in steady silence.

Without a single center to hold everything, texts were kept wherever there was space and care. A teacher might store a few in a chest beneath a table, while another placed them on a high shelf out of reach. In these rooms, the work of reading and teaching continued through habit and routine. The movement of students and scholars brought new copies into these places, and the quiet rhythm of study carried on far from the harbor where many of the texts had first been gathered.

Pages Formed in Distant Workrooms

Many of the surviving texts had been copied years earlier in workrooms where scribes sat side by side at long tables. Papyrus sheets were cut and pressed flat before being joined into long rolls. Ink was mixed from soot and plant resin, then tested on scraps to make sure it flowed evenly. Each line was written slowly, the scribe pausing often to check the spacing and shape of the letters before moving to the next word.

The work was steady and careful. One person might read from an older copy while another followed the sound, setting each word down in fresh ink. When a scroll was finished, it was dried, rolled tightly, and tied with cord to keep it from coming loose. Some were wrapped in cloth before being stored, protecting them as they were carried from place to place. Over time, these repeated efforts formed new copies that preserved the same words through patient labor and steady hands.

First Glimpses in Distant Hands

When these scrolls arrived in new cities, teachers and students gathered close to see them for the first time. A bundle would be opened slowly, and the writing examined with careful attention. Some recognized familiar passages and felt a quiet sense of relief. Others traced the letters with their eyes, unsure if the words matched what they had learned from older copies. The arrival of each text brought a mix of interest and hesitation as readers tried to place it within what they already knew.

Fellow scholars often studied the scrolls in silence before speaking. A few compared them with the versions they already held, noting small changes in wording or missing lines. Conversations stayed low, shaped by a shared sense that these texts had traveled far and might be the only copies left. The condition of the papyrus, worn at the edges and darkened with age, added to the concern that each one might carry damage that could not be repaired.

Watchful Attention in New Cities

As these scrolls began to appear in study rooms and small schools, their presence drew notice from those who oversaw public life. In some places, local leaders asked where the texts had come from and who had brought them. Visitors paused at doorways, listening as passages were read aloud inside. The interest was steady, and the questions were careful, shaping how openly the texts were handled.

Among scholars, concern grew in quiet conversations. Some wondered if certain writings might draw unwanted attention if read too often in public. A few chose to keep their scrolls wrapped and stored between lessons, bringing them out only when needed. Even among trusted peers, there was hesitation about sharing every copy that arrived. The growing awareness that these texts carried the weight of the past made their presence feel both valuable and uncertain at the same time.

Careful Handling and Quiet Limits

Scholars worked to study the scrolls with steady attention, often comparing one copy with another to check the wording. A reader would speak each line aloud while another followed the text, marking places where letters had faded or where a line seemed incomplete. Notes were written along the margins to guide future readers. In some rooms, the scrolls were placed on tables only when needed, then wrapped again and returned to storage.

Later stories claim that certain texts were tightly controlled or hidden, but clear records are limited. What can be traced comes from letters and notes that mention small steps taken to protect the scrolls from damage or loss. Some teachers allowed only a few students to handle them, and in a few cases, copies were made to reduce the need to use the originals. The care given to each text shaped how often it was opened and who could read from it.

The Burden of Guarding the Words

Those who kept the scrolls safe often carried the strain in quiet ways. Some teachers felt the weight of responsibility each time they opened a bundle, knowing the text might be one of the few left. They spent long hours checking for damage, wrapping and unwrapping the fragile rolls, and watching closely as students handled them. The care took time away from other work, and the worry followed them even after the rooms grew quiet.

In some places, the role of protecting these writings set people apart. A few were known in their communities for keeping old texts, and the attention brought unease. Neighbors asked questions, sometimes with curiosity, sometimes with doubt. Invitations to share lessons grew fewer in certain areas, and some chose to work in private rather than draw notice. The effort to preserve the scrolls rested heavily on those who kept them close, shaping their days and the spaces where they lived.

Quiet Custody and Formal Recordkeeping

In some cities, schools and local authorities began to keep simple records of the scrolls in their care. Lists were written to note which texts were stored in certain rooms and who was responsible for them. When a copy was borrowed for study, its movement was marked so it could be returned safely. A few institutions placed the most fragile pieces in chests lined with cloth, limiting how often they were opened and who could handle them.

Over time, the handling of these writings followed steady routines. Some texts were moved from private homes to larger rooms where they could be watched more closely. In a few cases, copies were made and sent to other places so the words would not be lost if one collection was damaged. The movement was quiet and careful, guided by simple rules meant to keep each scroll safe as it passed from one set of hands to another.

Years Measured in Fragile Pages

As decades passed, the scrolls that had been carried from place to place became older and more delicate. Some could no longer be opened without risk, and their contents were known only through copies made years earlier. Teachers relied on these later versions, reading from them as the original rolls remained wrapped and stored away. The effort to preserve the words shaped how knowledge was passed on, often through secondhand pages rather than the first written lines.

In time, questions grew about what might have been lost along the way. Small gaps appeared where sections were missing or too faded to read. Some copies showed slight changes in wording, and no one could be certain when those changes had taken place. The longer the years stretched on, the more the uncertainty settled into the work of those who studied the texts, leaving them to wonder how much of the past still rested unseen inside the fragile rolls.

Shelved Echoes of Distant Hands

Today, a number of these scrolls rest in climate-controlled archives, stored in long drawers or sealed cases lined with soft material. Their surfaces show cracks where the papyrus has grown brittle, and repairs made long ago are still visible as thin strips pressed across weak spots. Labels written in careful script mark where each piece was found or how it traveled from one city to another. Some are too fragile to open fully, so their contents are studied through earlier copies and preserved notes.

Archivists handle them with slow, steady movements, turning sections only when needed and recording every mark that can still be seen. Faded letters appear beneath soft light, and the texture of the fibers shows the labor of those who first made them. Each roll carries signs of its long journey, passed through many rooms and guarded by many hands. In the quiet of the storage rooms, the old pages remain, holding the faint memory of voices that once read them aloud.

Chapter 17
The Rooms That Fell Silent — Learning After the Loss

Dust settles slowly in a narrow study room where a teacher stands beside an empty shelf. A few scrolls remain, tied with worn cord and stacked in uneven piles, but wide spaces between them show where others once rested. Light from a high window cuts across the wall, revealing faint marks where shelves had been moved or removed long ago. The room is quiet except for the soft sound of pages being turned by a single student seated at a low table.

Outside, the street carries the noise of daily life, but inside the air feels still and heavy. The teacher lifts one of the remaining scrolls and sets it down with care, glancing toward the empty spaces as if measuring what is no longer there. Each lesson now depends on what has survived, and each page is handled with steady hands. The room holds a memory of what once filled it, a sense of voices and movement that have faded, leaving only a few fragile texts to carry the work forward.

Smaller Schools in Quiet Corners

In the years after the great collection was gone, learning moved into modest rooms scattered through the city. Teachers worked from small houses or narrow spaces near courtyards, gathering a few students at a time. Shelves held only a handful of scrolls, each one kept close and handled with care. The steady flow of people who once came to study in larger halls was replaced by smaller groups meeting in quieter places.

These rooms became the new centers of learning, shaped by what remained and what could still be taught. Students sat close to the teacher, listening as each line was read aloud. Lessons moved slowly, guided by the limited texts that had survived. The work continued through routine and habit, with each room carrying its own small share of knowledge. Life in the city moved on outside, while inside these spaces the effort to teach and learn went on in steady silence.

Fresh Copies from Fading Originals

In these smaller rooms, teachers and scribes worked to make new copies from the few scrolls that remained. Sheets of papyrus were laid out across low tables, trimmed to size, and pressed flat before writing began. Ink was mixed in small bowls, and the first lines were written slowly to set the spacing. Each letter was shaped with care, the scribe pausing often to check the fading words of the older copy beside them.

The work took many hours and often stretched across several days. One person might read from the worn original while another followed the sound, writing each line into place. When a page was filled, it was set aside to dry, then joined with others to form a new roll. These fresh copies carried forward what could still be read, shaped by steady hands working close to the last surviving texts.

First Readers in Narrow Rooms

When the new copies were finished, they were read aloud in small gatherings of students and teachers. The sound of fresh inked lines filled the quiet rooms as listeners leaned closer to hear the words that had nearly been lost. Some followed along with older fragments, checking the new text against what they remembered. A few nodded as familiar passages returned to life, while others paused at places where the lines seemed thinner or harder to follow.

Among the students, there was a quiet sense of uncertainty. They knew the copies came from texts that had already begun to fade, and no one could be sure if every word had been carried over. Teachers watched closely as the readings went on, listening for small mistakes or missing phrases. Each new roll brought a mix of relief and concern, as if the past had been pulled forward but not without a few places where the meaning felt fragile.

Watchful Eyes in the Learning Circles

As the new copies began to appear in more study rooms, their presence drew attention from those who guided public life in the city. Some leaders asked where the texts had come from and who was making the copies. Visitors stood near the doorways during lessons, listening as passages were read aloud. Their questions were calm but steady, shaping how openly teachers spoke about the work.

Among fellow scholars, the growing interest brought a sense of strain. A few worried that the careful copying of older texts might invite questions they could not easily answer. Conversations that once took place in open rooms moved into quieter corners. Some teachers hesitated before sharing a new copy, unsure how it would be received. The attention, though quiet, followed the work and settled into the spaces where the scrolls were read.

Reading Line by Line Under Careful Watch

Teachers and scribes worked slowly through the new copies, checking each line against the fading originals. A reader would speak the words aloud while another followed the text, marking places where letters were faint or where a line seemed broken. Notes were added along the edges to guide future readers. Some pages were copied more than once, as if repeating the work might help keep the words steady and clear.

Later stories claimed that certain texts were closely guarded or hidden, but surviving notes mainly show careful handling rather than strict control. Some teachers limited how often the oldest scrolls were opened, fearing they might tear. A few copies were stored in chests between lessons and brought out only when needed. These small steps shaped how the texts were studied and passed along, guided by caution and the need to protect what remained.

The Quiet Weight of Keeping the Work

Those who spent their days copying and teaching from the last remaining texts felt the strain settle into their lives. Hours bent over low tables left their backs stiff and their eyes tired, yet the work could not be rushed. Some spoke less with friends than they once had, giving more of their time to the careful task of preserving each line. The sense that the words might vanish if left alone followed them even when they stepped outside the study rooms.

In the city, their work set them apart. A few neighbors looked on with doubt, unsure why so much effort was spent on fading pages. Invitations to join gatherings grew fewer for some, and conversations often turned cautious when their work was mentioned. Even within their own circles, tension appeared as the burden of keeping the texts safe rested on only a few hands. The effort shaped their days, pressing quietly against the spaces where they lived and worked.

Rules Set in Quiet Offices

Over time, local schools and learning houses began to set simple rules for how the remaining texts were handled. Lists were kept to record which scrolls were stored in each room and who was allowed to use them. Some texts were moved to stronger storage chests, lined with cloth to protect them from dust and damp air. Teachers were asked to note when a copy was taken out and when it was returned, so the movement of each roll could be tracked.

In certain places, older scrolls were transferred from private homes into shared study rooms where they could be watched more closely. Access was sometimes limited to set hours, and only a few trusted hands were allowed to handle the most fragile pieces. These quiet routines shaped how the texts were kept, passed from one room to another through careful recordkeeping and steady attention to their condition.

Years Marked by Fading Voices

As decades passed, the copies made in those small rooms became the main source for teaching. Students learned from texts that had already been copied many times, each version carrying small changes that were hard to trace. Some teachers kept older fragments beside the newer rolls, reading from both to fill in missing lines. The effort to hold onto the words shaped how lessons were given, built on what could still be read and remembered.

Over time, gaps appeared that no one could fully explain. A few passages stopped in the middle of a thought, and no older version could be found to complete them. Names appeared in one copy but not in another, leaving quiet questions behind. The longer the years moved on, the more the work depended on pieces that had survived by chance, and the concern lingered that more might have been lost without anyone knowing when it slipped away.

Pages That Still Bear the Marks

Today, some of the later copies made in those smaller rooms are kept in archives, their surfaces worn from years of careful use. The papyrus has darkened and curled at the edges, and small repairs can be seen where tears were mended long ago. Faint notes written in the margins show where a teacher once corrected a word or filled in a missing line. These marks reveal the steady effort to keep the texts readable as time passed.

Archivists now handle the pages with slow, careful movements, placing them under soft light to study the writing without causing harm. Each sheet is stored flat and protected from air and moisture, held in cases that keep the fragile fibers from breaking apart. The lines written by unknown hands remain visible, shaped by the quiet work of those who tried to keep the words alive. In the stillness of the archive, the pages rest, carrying the memory of voices that once filled those silent rooms.

Chapter 18
The Voices That Remembered — Stories Passed by Word of Mouth

In a quiet courtyard at dusk, a small group sits close together as an older teacher speaks from memory. No scroll rests in his hands. Instead, he watches the faces around him, speaking slowly and choosing each word with care. A young student leans forward, repeating a line under his breath to fix it in his mind. The sound of distant footsteps echoes beyond the walls, but inside the courtyard the voices stay low and steady.

The teacher pauses often, closing his eyes as if searching for the next line. A second voice joins in, correcting a phrase, then falling silent again. What they carry is not written but held in memory, shaped by years of repetition. The night air cools as the lesson continues, each sentence spoken and repeated so it will not be lost. In the fading light, the words move from one voice to another, carried forward by those who remember them.

Lessons Carried in Open Courtyards

In many parts of the city, teaching no longer depended only on written texts. Groups gathered in open courtyards, along shaded walkways, or in the corner of a quiet room where a teacher spoke and students listened closely. Without enough scrolls to share, the spoken lesson became the center of learning. Students sat in small circles, repeating each line until it stayed clear in their minds. The sound of voices replaced the rustle of turning pages.

These gatherings followed steady routines. Lessons were held at certain hours, and students returned day after day to hear the same words spoken again. Older students helped guide the younger ones, repeating what they had learned to keep the lines strong in memory. In these spaces, learning moved through listening and speaking, shaped by the steady presence of those who held the words in their voices.

Words Set Down After Long Repetition

In some places, teachers began to write down what had been carried only by voice. Sheets of papyrus were placed on low tables, and ink was mixed carefully before the first lines were formed. A teacher would speak slowly while a scribe listened, setting each word into place as it was heard. The work moved at a steady pace, guided by memory and repeated checks to make sure the lines matched what had been spoken for years.

These new writings were shaped by careful labor. A phrase might be spoken more than once before it was written, and the scribe would pause to correct a letter or adjust the spacing. When a sheet was filled, it was laid aside to dry before being joined with others. Each new scroll grew from spoken lessons, formed by the effort to capture words that had lived in memory long before they were written down again.

First Readers of the Written Lessons

When the newly written scrolls were first shared, small groups gathered to hear them read aloud. Students who had learned the lessons by voice listened closely, comparing each line to what they remembered. Some nodded when the words matched the rhythm they knew. Others leaned forward, unsure when a phrase sounded slightly different from what they had heard in earlier gatherings. The room stayed quiet as each line was spoken.

Teachers listened with the same careful attention. A few paused the reading to ask that a line be repeated, searching their memory for the older form of the words. Some spoke softly to one another, wondering if time had shaped the lessons in ways they could not fully trace. The written text brought relief, yet it also carried a quiet concern that the voice and the page might not always hold the same sound.

Questions in the Circles of Learning

As the written versions began to appear in more study rooms, their presence drew steady attention. Some local leaders and teachers from nearby schools came to listen during the readings. They asked where the words had come from and how the lines had been remembered for so long. Their voices were calm, but the questions lingered in the air. The gatherings, once simple and quiet, began to feel watched in ways they had not before.

Among fellow teachers, unease slowly grew. Some wondered if putting spoken lessons onto scrolls might change how the words were understood. A few spoke in private about whether certain lines should be shared more carefully. The act of writing what had long been carried in memory drew notice, and with it came a sense that each lesson now held a weight that others were beginning to measure.

Reading Aloud to Hold the Words Steady

Teachers worked carefully to study the new scrolls beside the spoken lessons they had known for years. One person would read from the written page while another repeated the line from memory, listening for any change in sound or order. Small notes were added along the edges where a phrase seemed uncertain. In some rooms, the same passage was read more than once, with pauses between each reading to make sure the words stayed clear.

Later stories suggest that certain lessons were kept from being written down, but surviving notes mostly show careful checking rather than strict limits. Some teachers allowed only a few students to copy the lines, fearing that mistakes might spread if too many versions were made. In a few places, the newest scrolls were stored away after lessons and brought out only when needed. The work moved slowly, shaped by the effort to keep the spoken and written words close together.

The Strain of Carrying Living Words

Those who led the spoken lessons felt a new pressure settle into their daily lives once the words began to be written down. Some worried that a mistake made in memory might now be fixed on the page for others to see. They spent long hours repeating lines alone, testing their recall before speaking in front of students. The effort left them tired and quiet, their work following them into the late hours of the night.

Among their peers, distance began to grow. A few teachers questioned whether the spoken lessons should have been written at all, and conversations turned careful when the subject came up. Some who once worked side by side found themselves standing apart, unsure of how their roles would change. Even within small learning circles, tension appeared as each person carried the weight of keeping the words right, knowing that others were listening closely for every line.

New Routines for Spoken and Written Lessons

Over time, small schools and teaching circles began to set simple rules for how spoken lessons and written copies were used together. Teachers noted which lines had been written down and which were still carried only in memory. In some places, scrolls were kept in certain rooms and brought out only at set times. Students were asked to listen first before making their own copies, so the words would be heard clearly before they were written.

In a few learning houses, lists were kept to track who was responsible for each scroll and where it was stored after lessons ended. Some copies were moved from one place to another to keep them safe, passed quietly between trusted hands. The work of speaking, writing, and storing the lessons followed steady patterns shaped by these small routines. The words moved through rooms and voices, guided by careful handling and quiet recordkeeping.

Echoes Carried Across Generations

Years later, the lessons once spoken in quiet courtyards could still be heard in distant places, passed from teacher to student through careful repetition. Some of the lines that had first been set down in fresh ink were now read from copies made long after the original writers were gone. Voices carried the words forward, shaped by habit and memory. In some rooms, the spoken form remained strong, while in others the written version guided the lesson.

As time moved on, differences began to appear between what was spoken and what was read from the page. A phrase remembered in one place did not always match the version written in another. Teachers paused at certain lines, unsure which form had come first. The uncertainty settled quietly into the work, carried in the space between voice and ink, where no one could fully say how the earliest words had once sounded.

Faint Marks Where Voices Became Ink

In archives today, some of the earliest written versions of those spoken lessons remain, their surfaces marked by careful hands that tried to capture each remembered line. The writing is uneven in places, showing where a scribe paused or corrected a word. Small notes in the margins suggest that a teacher once checked the phrasing against what was spoken aloud. The pages carry signs of long use, their edges worn where they were unrolled and rolled again.

Archivists now store these sheets in quiet rooms where light and air are controlled to protect the fragile fibers. Each mark is studied closely, and the faint corrections are recorded before they fade further. The writing stands as a trace of voices that once filled open courtyards, carried from memory to page. In the stillness of the archive, the lines remain, holding the sound of lessons that were once spoken into the evening air.

Chapter 19
The Names That Remained – Remembering the Keepers of Knowledge

A narrow room lit by morning light holds a long table covered with worn pages. A reader bends over a scroll, tracing a faded line with his finger until he finds a name written in the margin. He pauses, then reads it aloud, the sound quiet but steady. Another listener leans closer, repeating the name under his breath as if testing its weight. Outside, footsteps pass along the street, but inside the room the focus stays on the writing.

More pages are turned, each carrying small notes written by hands from long ago. Some names appear only once, others return again and again beside the same lines of text. A few are marked with brief remarks, simple signs left by those who copied or protected the words. The room fills with the soft rustle of turning pages as each trace is found and read. In the still air, the names remain, carried forward by the work of those who continue to speak them.

Small Circles That Held the Records

In the years that followed the loss of the great collection, groups of teachers and scribes kept their work alive in modest rooms across the city. These places were often tucked behind busy streets, known mainly to students and those who came to learn. Tables were set close to the walls, and shelves held the few texts that had survived. The people who cared for them worked in steady patterns, gathering each day to read, copy, and teach what they could.

Within these small circles, certain names were spoken often. They belonged to those who had once protected the scrolls, carried them through danger, or spent years copying fading lines. Students learned these names along with the lessons, repeating them as part of the work. In quiet rooms filled with low voices and steady light, memory became part of the setting, shaped by the people who continued to keep the record alive.

Names Written into the Margins

As new copies were made, scribes sometimes added small notes beside the main lines of text. These markings were written in thinner ink, placed carefully in the open space along the edges. A name might be set beside a passage to show who had copied it or who had kept the original safe. The work was done slowly, with each letter shaped to fit within the narrow margin without touching the main writing.

The process required steady hands and close attention. A scribe would finish the main lines first, then return to add the small notes at the side. Some names were written again and again across different pages, carried forward as new copies were made. Over time, these small additions formed a quiet record of the people who had handled the texts, their presence marked in ink along the edges of the page.

First Eyes on the Written Names

When readers began to notice the names written along the margins, they paused to study them closely. Some leaned nearer to the page, tracing each letter with care as they tried to sound the name aloud. A few recognized certain names from lessons they had heard before, spoken by teachers as part of the story of how the texts had survived. Others were less certain, unsure whether the marks had been made at the same time as the main writing or added later.

Among students and teachers, the discovery stirred quiet conversation. Some wondered how many hands had carried the texts before they reached this room. A few spoke softly about the responsibility of copying the names exactly as they appeared, fearing that even a small change might erase the memory of someone who had worked to protect the words. The pages held more than lessons now, and the added names brought a sense of weight to each reading.

Unease Around the Written Record

As the names written along the margins began to draw notice, questions surfaced among those who oversaw learning circles and local schools. Some wondered why certain individuals were being recorded while others were not. Visitors who came to hear readings sometimes asked who the names belonged to and how they had come to be added beside the text. The attention was steady and quiet, yet it lingered in the room as each page was turned.

Among fellow scribes and teachers, a sense of strain began to grow. A few worried that writing names might invite closer scrutiny from those outside their circles. Others spoke carefully about whether every name should be copied again or left only where it first appeared. The added markings, once a simple way to remember, now carried a weight that could not be easily set aside.

Careful Reading of the Margins

Teachers and scribes began to study the names with the same care they gave to the main lines of text. Some compared one copy with another, looking to see if the same names appeared in the same places. A few made small lists to keep track of which names were found on certain pages. The work moved slowly, shaped by the need to copy each letter exactly as it was written, without change.

Later stories suggest that certain names were hidden or removed, but clear records mostly show careful handling rather than strict control. In some rooms, teachers chose to read the main text aloud without speaking the names in the margins. A few copies were kept closed except when needed for study. These steps guided how the pages were used, as the markings were treated with the same care as the words beside them.

The Weight of Being Named

For those whose names appeared along the margins, the recognition carried a quiet strain. Students sometimes spoke the names aloud during lessons, and the sound of them lingered in the room. A few of the living who shared those names felt uneasy when they heard them repeated, unsure how their connection to the work might be seen. The simple act of writing a name beside the text could draw notice that followed a person through daily life.

Among teachers and scribes, tension grew in small ways. Some worried about adding their own names to a page, unsure how long the writing might last or who might read it later. Others felt the weight of copying a name wrong, knowing that even a small mistake might change how someone was remembered. Conversations grew careful, and a few chose to work more quietly, stepping back from gatherings where the pages were read aloud.

Quiet Practices Around the Written Names

Over time, small learning houses began to set simple practices for how pages marked with names were handled. Some teachers chose to store those copies in separate chests, keeping them wrapped and opened only during certain lessons. Lists were kept to note where each marked scroll was placed and who had used it last. The movement of these pages was watched more closely, shaped by a steady effort to keep them from being damaged or misplaced.

In a few places, copies that carried many names were moved from private rooms into shared storage spaces where they could be kept under watch. Teachers recorded when a scroll was taken out and when it was returned, and some asked that only certain students handle the older pages. These quiet routines guided how the marked texts were passed from one place to another, handled with care and kept within small circles where their use could be tracked.

Names Carried Across the Years

Decades later, the names written along the margins continued to appear in new copies made from the older pages. Students who had once learned from those marked texts repeated the names as part of their lessons, carrying them into other cities and schools. Some names remained clear and were written again and again, while others grew faint as copies were made from copies. The act of remembering became tied to the written marks, passed forward through steady repetition.

Over time, questions began to settle around which names belonged to the earliest hands and which had been added later. A few copies showed different spellings, and no one could be certain when those changes first appeared. Some names stopped appearing in newer texts, leaving only faint traces in the oldest pages. The years carried the memory forward, yet the distance between the present and the first written lines left space where certainty slowly faded.

Faded Signatures in Quiet Archives

Today, the marked pages that carry those names rest in carefully controlled archives, protected from light and damp air. The ink has thinned in places, and some letters have faded until they can be seen only under close study. Archivists use soft brushes to clear away dust and handle the sheets with steady hands, turning them slowly to keep the fragile fibers from breaking. Small notes written long ago remain along the margins, each one a trace of a person who once held the page.

Scholars examine these markings under gentle light, comparing one copy with another to follow the path of each name. Some appear again and again across different pages, while others surface only once before fading from the record. The quiet rooms where the texts now rest hold these traces in stillness. In the thin lines of ink, the presence of those who kept the words alive remains, resting at the edge of the page where their hands once moved.

Chapter 20
Epilogue:
The Rooms That Still Hold Them — Silence, Waiting, and the Careful Hand

In many cities across centuries, there were moments when doors were closed not to end learning, but to contain it. In monasteries, temples, and private study houses, shelves were moved and chests were sealed. A text once read openly was placed in a corner, wrapped in cloth, and handled only when needed. The decision was rarely announced. It showed itself in small changes, in how often a scroll was opened and how carefully it was stored.

Records from different places describe these choices. In some regions, teachers gathered texts and kept them in locked rooms rather than burning them. In others, leaders ordered that certain writings be placed out of public reach but not destroyed. The work of containment moved quietly, shaped by the belief that some knowledge carried risk if used without care. The rooms stayed closed, yet the texts remained inside, resting in silence.

Decisions Made in Quiet Offices

In official buildings, scribes wrote down the movement of texts with steady hands. Lists were kept to record where certain writings were stored and who was allowed to handle them. Some were placed in protected rooms that only a few could enter. These choices were not always written as strict laws. They appeared as routines, small rules followed day after day, shaping how knowledge was handled.

In one place, a text might be moved from a public hall into a smaller chamber. In another, it might be copied once and then set aside, its use reduced to rare moments. The aim was not always to remove it from existence. It was to limit how widely it moved. These actions repeated across time, in different lands and under different leaders, carried out through steady work rather than open command.

Monasteries That Chose to Lock the Doors

In the hills beyond busy towns, monasteries became places where texts were kept under steady watch. Shelves stood behind thick doors, and only certain brothers were given the keys. Some writings were still copied by hand, yet others were placed in storage and opened only on rare occasions. Records show that this was done not to destroy the pages, but to keep them from being read without care.

Over time, these quiet rules shaped daily life in the monasteries. A book that once rested on an open table might be wrapped and placed in a wooden chest. Notes were written to record where it had been stored, and the chest was moved to a safer room. Years passed, and the work of copying shifted to other texts. The older writings remained in place, held in silence by those who believed they should be kept but not spread.

Rooms Sealed During Times of Unrest

When cities faced danger, institutions often chose to protect their collections by hiding them. In times of war or sudden change, books and scrolls were gathered and placed in cellars or behind stone walls. Doors were sealed, and the rooms were left closed for years. The aim was not to erase what was inside, but to keep it safe until calm returned.

In some cases, those who sealed the rooms never came back to open them. Generations passed, and the memory of what had been stored faded. Later records mention the discovery of sealed chambers filled with writings that had been placed there long before. The pages were still wrapped, their edges marked by age. The decision to hide them carried the texts across time, even when the people who made that choice were long gone.

Traces Left After Use Ended

When active reading stopped, the texts did not always disappear. They stayed in the places where they had been stored, waiting in dark rooms and sealed containers. Some were found years later, still wrapped in the same cloth. Others remained listed in old records long after no one had opened them.

In archives today, these traces can still be seen. Labels show where a writing once rested and when it was last handled. Some pages bear marks from long ago, small notes added by someone who read them before the doors were closed. These signs tell of a time when the text was part of daily life, then slowly set aside. The shift from use to storage happened step by step, guided by careful choices rather than sudden acts.

Patterns That Returned Across Time

Across different cultures and centuries, similar moments appeared. A body of knowledge would grow, draw attention, and then be handled with greater care. Access would narrow. A text once shared widely would be passed only between trusted hands. The pattern repeated in many places, shaped by the same human sense of caution.

Some institutions chose to preserve certain writings even when they no longer used them. Others placed them in collections where they could be studied in limited ways. The hesitation that followed did not always come from fear alone. It came from a sense that once words were released, they could not be called back. So the choice was made to slow their movement, to hold them in place and let time settle over them.

Archives That Watched in Silence

In modern times, the role of the archive took on the same quiet pattern. Texts that were no longer part of daily study were placed in storage under careful watch. Rooms were built to control light and air. Shelves were arranged to keep each piece safe from harm. The work of preservation became a steady task, carried out with patience and care.

Many of these texts had not been used for years, yet they remained part of the collection. Their presence was recorded, their movement tracked, and their condition checked. Some were opened only once in a great while. Others were left closed, resting in place. The silence that surrounded them became part of their story, a sign that they had been kept safe by being left alone.

What Remained When the Doors Closed

When scholars stepped away and institutions limited access, the texts stayed where they had been placed. The rooms grew quiet, and the pages rested in stillness. Dust gathered, and the ink aged, yet the words remained. The decision to hold back knowledge left behind a different kind of record, one written in silence rather than speech.

In archives and storage rooms today, these writings still rest on their shelves. Their pages show the marks of many hands, the signs of years spent in careful keeping. The pattern has not disappeared. It has simply shifted form, carried forward by those who choose to protect what they hold. Beyond the closed doors, the world continues to move. Inside, the texts remain where they were left, waiting in the quiet, held in place by patience and time.

Made in the USA
Monee, IL
04 November 2026

67384793R00089